Microwave Materials
for Wireless Applications

For a complete listing of titles in the
Artech House Microwave Library,
turn to the back of this book.

Microwave Materials for Wireless Applications

David B. Cruickshank

ARTECH
HOUSE

BOSTON | LONDON
artechhouse.com

Library of Congress Cataloging-in-Publication Data
A catalog record for this book is available from the U.S. Library of Congress.

British Library Cataloguing in Publication Data
A catalogue record for this book is available from the British Library.

Cover design by Adam Renvoize

ISBN 13: 978-1-60807-092-3

Contents

Preface

The purpose of this book is to provide a bridge between the engineers, scientists, and students of microwave and RF engineering involved in passive devices, and the corresponding group who design and manufacture materials for them, including managers directly or indirectly involved in the manufacture and use of these materials. For the former, the intention is to give a more systematic understanding of the chemistry and physics of the materials they use in order for them to make the best possible choice for their application. For the latter, the aim is to provide some insight into the microwave world so they, in turn, can provide a better range of choices.

The treatment is intended to be nonmathematical wherever possible, but some knowledge of chemistry and microwave engineering is assumed.

The book is aimed particularly at engineers designing magnetic- and dielectric-based components like ferrite isolators and circulators, dielectric based filters, and antennas and radomes where there are significant specialist materials requirements.

Because of the emergence of tunable device applications, a section, Chapter 11, on the potential tuning attributes of ferrimagnetic, dielectric, and paraelectric materials is presented.

Chapters 1 through 7 are concerned with the basic characteristics of magnetic, dielectric, absorbing materials, and metals with respect to their microwave properties. Chapters 8 through 11 are devoted to device applications, and Chapter 12 covers materials measurement.

The history of magnetic materials began with spinel ferrites, but the discovery of ferromagnetic garnets began a period of intense activity through to the late 1970s. By then it was thought most aspects of microwave ferrite devices were known. However, the emergence of cellular radio in the 1980s through to today means that more of these devices than ever are being produced, to ever more exacting requirements for insertion loss, temperature stability, power handling, and intermodulation characteristics. The body of information generated in the 1970s is being reinvestigated and expanded to explain and improve today's materials and devices, but suffers from a consequent generation gap in experience, which I hope this book will help to fill.

Dielectric materials, by contrast, did not become a topic of major interest for microwave engineers until the cellular revolution, and there is a much more continuous thread of development of plastic composites, laminates, and ceramic materials from then until now. It seems likely the continuous pressure to reduce size in wireless infrastructure will continue to fund the search for the impossible: a high dielectric constant, high Q material that is temperature-stable, cheap, and readily available.

Acknowledgments

I have been extremely fortunate to have been exposed to some of the best people in the field of microwave materials, on both sides of the materials and microwave engineering fence.

From my first days in this field at Ferranti Radar in Edinburgh, I am grateful to Phil Cohen, Bob Murray, and Bill Hepburn for first infecting me with enthusiasm for this topic. At MESL, I was privileged to know Robbie McLean, Fred Aitken, and Professors Jeff Collins and Joe Helszajn, and later at Racal MESL, Dr Bill Alton, Dr Terry Nisbet, Ian Alexander, Tom Wood, and George Simpson. From my early days at Trans Tech, among the many people there I acknowledge the late Al Blankenship and the late Russel West. In more recent times at Skyworks Trans Tech I am grateful to Bob Huntt, Dr. Tyke Negus, Elwood Hokanson, John Deriso, Dan Tipsord, and most of all Dr. Mike Hill.

Academic influence over the years came from Professor Tony West and Dr. Karen Trotman (my daughter) at Aberdeen University, Professor Tony Pointon and Dr. David Nixon at Portsmouth University, Dr. Doug Adam and later Dr. David G. M. Cruickshank (my son) at my alma mater, Edinburgh University, and Professor Lionel Davis and later Dr. Tara Miura at UMIST. Of the many people at MMA gatherings, I acknowledge particularly Professors Bob Freer and Jerzy Krupka.

Finally, I am grateful to my son Simon, and particularly to my wife Ling for her patience and support while writing this book.

Introduction

This book is intended to allow the reader to begin to see, as the chemistry and physics of key materials is explained, how each element in a material can influence the behavior of the material, and how common strands appear between apparently unrelated material types. More than half of the 90 or so potentially usable elements in the periodic table appear, often in different roles, and the aim is to make the reader sufficiently familiar with their chemistry to anticipate their likely role in forming the structure and influencing the behavior of the material they inhabit. In recent times the cost and availability of certain elements has become an issue, and this will be addressed as each material class is considered.

The majority of the book is about bulk materials, although a number of thick and thin film applications are discussed. The tendency in the wireless world is towards miniaturization, but there are some fundamentally inescapable limitations. The first is frequency, since the lowest frequency determines the greatest wavelength, and therefore resonator or circuit wavelength minimum dimensions are limited by the highest dielectric constant and permeability available consistent with low losses. At the highest frequencies, attractive for available bandwidth, range becomes a limitation because of atmospheric attenuation, and propagation of radio signals is confounded by reflection and multipath effects. As a result, most outdoor wireless is further limited to where it is already most crowded, in the 400-MHz to 3-GHz frequency range. Second, when we look at the type of magnetic or dielectric resonator that can be used, the smallest part that can be realized is a relatively large TE, TM, or TEM mode resonator or circuit element. As we shall see, the highest dielectric constant available at microwave frequencies with reasonable losses and temperature stability is less than 100, and the highest permeability less than 10. As a result, most resonator-based passive devices, other than acoustic resonators, are in millimeter or centimeter sizes, not the micron scale of semiconductor or related thin film devices. Within the 400-MHz to 3-GHz band, it was anticipated that spread spectrum and similar wideband methods would use large pieces of the spectrum to avoid excessive transmit/receive filtering and reduce power levels as receivers became ultrasensitive to power but indifferent to interference. In practice, available bandwidth is extremely small, typically a few megahertz, or at best a few tens of megahertz, because of preassigned spectra, multiple operators, and the need to keep options open for time division and frequency division multiplexing. If we add to that mix cositing of multiple services on the same base station, the result is a need for complex, narrowband, near-antenna filters, very low loss components and, because of multichannel systems, very low intermodulation. A resultant constant theme of the book is the need for very low dielectric losses in plastic or ceramic materials, low dielectric and magnetic losses in ferrites, and very high electrical conductivity metal structures. Outdoor exposure, relatively high power levels and crowded subsystem packaging have added temperature stability and thermal conductivity to these themes.

These themes are addressed sequentially, with the objective of guiding the reader logically through to the optimum choice of material for his or her application, whether it is a magnetic, plastic, ceramic dielectric, absorber, metal, or a composite material of any of these types. The applications themselves are then discussed with that background in mind, focusing on those most bulk materials intensive. The range of applications include the latest types of ferrite devices, resonators, filters, antennas, radomes, and tunable devices (currently an active area of new device development). This group of devices is generally considered "passive." This literally has come to mean microwave devices without semiconductors, and to a large extent this is the definition being followed here. In practice, the boundary is a little blurred by semiconductor switches sometimes being called passive, while, for example, many passive components and subsystems may contain semiconductor detectors, electronic drivers, and so on.

The range of frequencies covered by this book is from a few kilohertz to close to 100 GHz, if we exclude mention of combined microwave and infrared materials, which are discussed in Chapter 5. The reason for this is twofold. First, an understanding of the behavior of materials in the mainstream of the microwave wireless range is deepened by knowing what is going on either side of the electromagnetic spectrum. Second, although it can be argued that wireless will remain centered on the 400-MHz to 6-GHz range, there are exceptions like point-to-point radio, currently used up to about 90 GHz, as data rates demand more available bandwidth and where the corresponding shorter range is not an issue. In addition, at the other extreme, applications at 13.56 MHz, the industrial, scientific, and medical band, and other applications at VHF frequencies, are considered where they are natural extensions of microwave materials technology.

A table of common units (Table I.1) and their common abbreviations and conversion factors is shown next. Note that for the purposes of this book, permeability (μ, m, or μ_r are used), dielectric constant (ε or ε_r are used), and dielectric or magnetic loss tangents (tan$d$ or tan$m$) and their reciprocal quality factors (Q) are assumed to be dimensionless, and are omitted from the table.

Whether the reader is an experienced design or sourcing engineer, a materials engineer or manufacturer, or a student of any of these disciplines, the intention is to add to their knowledge of microwave materials.

Table I.1

Table of Units, Abbreviations, Symbols, Conversion Factors, and Coefficient Calculations

Unit	Abbreviation/Symbol	Conversion Factor/ Calculation
Meter	m	39.4 inches
Centimeter	cm	0.394 inch
Micron	μ	39.4 microinches
Angstrom	A	10^{-4} micron
Valency of element ion	(Chemical symbol) +n or $^{+n}$	n = Number of electrons lost or shared
Gram	g	0.03527 ounce
Watt (heat)	W or w	2.39 calories (cal)/sec
Degree Centigrade/ Kelvin	C/K	1.8 degrees Fahrenheit (°F)
Thermal conductivity	W/m.C	0.00239 cal./s.cm.C
Thermal expansion	cm/cm/C	1.8 inch/inch/°F
Watt (RF power)	W or w	30 dBm
Isolation/insertion or return loss	dB	Log_{10} relative power
Magnetic field strength, Oersted	Oe	$1,000/4\pi$ ampere turns/m
Magnetic induction, Gauss	G	10^{-4} Tesla
Temperature coefficient frequency, parts per million (ppm)	Tf or T_f (in ppm/C)	Frequency shift/ frequency/°C
Second-order temperature coefficient of frequency	T'f or T'_f (in ppm/C/C)	Frequency shift/ Frequency/°C/°C
Temperature coefficient of dielectric constant, r	Te or T_e (in ppm/C/C)	Change in Er/Er/°C

1

Garnets

1.1 Introduction

Garnet minerals occur in nature and have been used for many centuries as semiprecious stones (Figure 1.1). Laboratory synthesis began in earnest in the early 1950s, and soon the garnet structure was found to accommodate a wide variety of elements [1]. However, the compounds that were produced were of academic interest only.

The discovery of the ferrimagnetic properties of the garnet family in the late 1950s stirred a great deal of interest in the physics community, comparable to the discovery of oxide-based superconductors of the late 1980s and 1990s [2]. For almost two decades the pages of the magnetic section of the applied physics and IEEE journals were filled with articles about the properties and application of this group of materials based on yttrium iron garnet (YIG). The reason for this interest was the unique properties of the structure. Here was a compound where each of the three different sites (referred to as the dodecahedral or c site, the octahedral or a site, and the tetrahedral or d site) could be independently doped with a whole range of metal ions to test the

Figure 1.1   Garnet crystals found in nature.

individual properties and overall effect of each ion. It rapidly became similar to the fruit fly of geneticists—a single change of magnetic ion, instead of a gene, would produce measurably different results. A further bonus was that it was easy to make in single-crystal (Figure 1.2) and polycrystalline form, allowing the study of, for example, single ion

Figure 1.2   Synthetically grown YIG crystals.

magnetic anisotropy, or nonlinear behavior under varying RF power and magnetic fields. The spinel structure, with its vacant sites and ease of inversion (Chapter 2), made it difficult to pinpoint the location and properties of individual ions, and it rapidly took a back seat to this new family of materials.

Today, in the twenty-first century, we are still reaping the benefit of these decades of study. The insertion loss and intermodulation characteristics of modern devices can only be understood and improved with this legacy of information. The purpose of this chapter is to begin to examine this information and, in subsequent chapters, apply it to current materials and devices. We will look at each garnet crystallographic site in turn and discuss the characteristics of most of the known useful substitutions.

1.2 Garnet Structure and Chemistry

The dodecahedral or c site is filled with yttrium (Y) ions in YIG. The formula unit is actually (Y_3) (Fe_2) (Fe_3) O_{12} representing in brackets the contents of the c, d, and a sites. There are 8 of these formula units per unit cell, which is cubic in crystal classification, with a side of 12.377 Angstroms (Angstrom is a unit of length A, where 10^8 A = 1 meter). The unit cell dimensions change a little as we substitute for Y or iron (Fe). Using these and the formula weights of different compositions we can calculate the theoretical density, useful in evaluating the quality of polycrystalline materials. The structure can be seen in Figure 1.3.

To understand magnetic behavior in garnets, it is important to appreciate the properties of atoms and their valency as ions. When metal atoms lose electrons with a negative charge to become metal cations with positive charge, the positive charge, proportional to the number of electrons lost, is called the valency of the ion. The value is generally related to the outer electron energy levels and their availability to form compounds by losing or sharing electrons with other ions.

The valency of Fe ions in YIG is +3, as are yttrium ions and all lanthanides, although europium (Eu) and cerium (Ce) actually prefer +2 and +4, respectively. Lanthanides are a series of 14 similar atoms in the periodic table with useful magnetic properties when forming cations in oxide materials. They have very similar properties because an

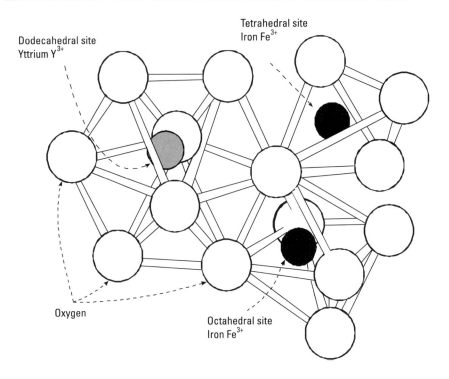

Figure 1.3 The structure of YIG. Note oxygen ions are in white, the dodecahedral Y ion is in gray, and tetrahedral and octahedral Fe ions are in black.

inner energy level, the 14f "shell" is being progressively filled through the series. They include all of the so-called rare earth elements, plus yttrium, which enter the dodecahedral site. Yttrium is not part of the lanthanide series but is included as a "rare earth" for convenience because it is found in the same ores and is very similar in its chemistry.

It should be appreciated that when nontrivalent ions are introduced, valency balance should ideally be maintained. Divalent calcium (Ca), for example, is used to balance tetravalent zirconium (Zr) on a one for one basis. Two Ca ions are necessary to compensate for pentavalent vanadium (V), and so on. Because ions like Fe, cobalt (Co), and manganese (Mn) can have multiple valencies, to get the desired magnetic effect from a particular valency, suitable compensation is required. The presence of multiple valencies of the same ion, for example, Fe, is normally highly undesirable.

1.3 Magnetism and Ferrimagnetism

Before discussing the behavior of ferrite materials, some basic concepts need to be considered about the origin of magnetism and how ferrimagnetism arises in crystal structures.

Electrons have two opposite spin states, which create magnetism at the atomic level. When spins are paired with their opposite, these cancel with no net spin. When unpaired these have spin and, because of quantum mechanical rules, several unpaired electrons can be present in a cation, creating even stronger magnetism. Examples are Fe+3 with 5 unpaired electrons, and Gd+3 with 7. The total resulting magnetism is expressed in Bohr magnetons at the atomic level.

Magnetism (symbol Ms) in ferrimagnetic materials occurs when all similar ions on the same crystallographic site add their atomic level magnetism to create a magnetic subsystem or sublattice. When more than one sublattice is present, with potentially different, opposing, directions of magnetization, this is defined as ferrimagnetism. The net magnetization is the vector sum of these sublattices, and is expressed in gauss (10 gauss = 1 milliTesla or mT). Values of magnetization are stated as the 4πMs of a given material in either gauss or mT.

Magnetocrystalline anisotropy is the angular relationship between the crystallographic direction and the direction of the magnetization vector of a given ion. It is expressed as a series of constants, K1, K2, and so forth, but for many purposes only K1 is used. The resonance line width can be approximated to K1/Ms. Spin canting is the influence of the crystallographic environment on a magnetic ion's spin or magnetization direction when different cations are introduced.

Magnetic resonance is derived from spinning electrons, which when excited by an appropriate radio frequency (RF) will show resonance proportional to an applied magnetic field and the frequency. At the molecular level this translates into cooperative behavior as the net magnetization and the gyromagnetic ratio express the relationship between the applied field and the frequency. The width of the resonance peak is usually defined at the half power points (3 dB of power on a log scale) and is referred to as the ferrimagnetic resonance line width. The applied field and hence line width is usually expressed in Oersteds (1 Oersted = $1,000/4\pi$ Ampere-turns/meter).

Before we look at the various site substitutions, we need to examine the validity of some of our assumptions about them. Ferrimagnetism is formally defined as a material with more than one magnetic sublattice, when the sublattices have opposing magnetizations. A sublattice is defined as the sum of the magnetization of all the magnetic ions in the same type of site, in this case either c, a, or d sites. These ions on the same type of site are physically (on an atomic scale) remote from each other, but "couple" to each other via oxygen ions. This effect is called super-exchange [3] and allows these magnetic spins (moment) to line up with each other. Hence they can be regarded as an array or "team" of ions called a sublattice, with a finite total magnetic moment or magnetization per unit cell. Because YIG has three tetrahedral ferric ions for every two octahedral, and the two sublattices oppose each other, the tetrahedral sublattice "wins" by one, and dominates the net magnetization at all temperatures. Because Y ions are nonmagnetic, they do not "play."

1.4 Magnetic Ions Behaving Badly

The magnetic behavior of the overall material is determined by the net magnetization sum or difference of its sublattices. Different ions, however, have vastly different magnetic anisotropy. In practice, ions used as substitutes may not behave predictably because of, for example, spin "canting" induced either by the magnetic ion itself or by the effect of nonmagnetic ions on the environment of adjacent magnetic ions, reducing the degree of alignment. Hence the assumption that the net magnetic behavior is the sum of independent sublattices or single ion anisotropy may not always apply in predicting magnetic properties.

The sublattice magnetization can be disrupted in a number of ways. Introducing a nonmagnetic ion into a magnetic sublattice will weaken the effect, reducing the overall magnetization of the sublattice. In the same way, progressively introducing a magnetic ion into a nonmagnetic sublattice having initially no super-exchange interaction will introduce magnetism into the sublattice, eventually creating a third magnetic sublattice, as in the case of YIG when nonmagnetic Y is progressively substituted by a magnetic lanthanide ion. In the case of c-site substitutions for Y, very unusual magnetic anisotropy and nonlinear behavior can result. We shall also see that the amount of substitution is limited in some cases. Beyond a certain limit, the ion will not enter its preferred lattice site and either remains on the "outside" in a second phase compound or leaks into another site. Ion size and crystallographic orientation preferences may compete at high substitution levels, or substituting ions are influenced by the ion size and coordination of ions on other sites.

1.5 Lanthanides and Dodecahedral Substitution

Although the entire 14f lanthanide series ions can be substituted and have been studied, in modern devices only a few are used. This is partly because many behave similarly, so choice is based partly on raw materials cost, and partly on their effectiveness as magnetic ions. Some, of course, are nonmagnetic and serve no purpose.

When choosing a lanthanide, there are two different objectives. In the case of gadolinium (Gd), we look for weak spin lattice coupling, or relaxation, to minimize anisotropy and hence magnetic losses, but a large magnetic moment relative to Fe. Gd has a moment of 7 Bohr magnetons versus 5 for Fe at zero degrees Kelvin. Because Gd is in the c-site sublattice and Fe is in the a and d site, and Gd loses its magnetic moment much faster than Fe as the temperature rises, we create a series of temperature compensation points, dependent on the concentration of Gd, where the net moment of all the sublattices is zero. We can then create a series of materials with a relatively flat rate of change of net moment with temperature, from the compensation point to the Curie temperature, using substitutions in the range 10 to 60% of Gd for Y. This can be seen in Figure 1.4.

In the case of holmium (Ho) and dysprosium (Dy), the objective is to find a strong or "fast" relaxer with strong spin lattice coupling, such that a relative small substitution or "doping" will have the desired effect without changing other aspects of the material's properties, such as magnetization. Spin lattice coupling is the mechanism that allows the relaxer ion to increase the peak power handing by deferring the onset of nonlinearity. Doping can range widely in value, but practical values from 0.1% to 1% of Ho for Y are used to balance peak power handling and insertion loss. A complete range of relaxers has been

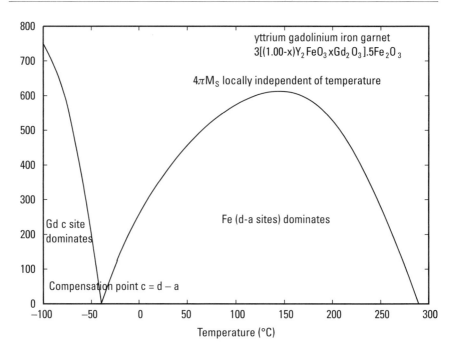

Figure 1.4   Schematic of gadolinium substitution in YIG.

described in detail [4], and a selection of the more common choices appears in tabular form in Table 1.1 in terms of their incremental effect on resonance line width, effective line width, and spinwave line width, for a fixed level of substitution. These characteristics will be described in more detail in Chapter 8.

Before the influence of these ions is considered further, a brief introduction to ferrite device parameters is necessary, although they are discussed in more detail in Chapter 8. The insertion loss of a device is the additional loss the device introduces into a microwave transmission line, which may be waveguide, coaxial cable, microstrip, triplate, and so forth. The loss is the sum of the reflected and transmitted microwaves, and includes a contribution from the ferrite's dielectric and magnetic losses. Bandwidth is the ratio of the operating frequency range of the device divided by the center frequency. At high power, losses may increase, and this is described as nonlinear behavior. Intermodulation is a nonlinear or mixing effect in a material, which produces a series of

Table 1.1
Relative Effectiveness of Relaxers in YIG

Relaxer Effect on YIG at 0.01 mole Substitution	Increase in 3-dB Resonance Line Width (Oe)	Increase in Effective Line Width (Oe)	Increase in Spinwave Line Width (Oe)
Sm +3	14.1	14.7	5.3
Gd+3	Not measurable	0.7	0.2
Tb+3	104	102	29.3
Dy+3	26	27.2	8.5
Ho+3	46	53	18.2
Co+2	See 1.6.3.	10.5	4.2

Source: [4].

sum and differences frequencies if two or more single frequencies are applied to the device simultaneously.

The three different types of line width in Table 1.1 require separate consideration depending on the application. Effective line width is a relative indication of the dopant ion's effect on insertion loss in a below-resonance device, whereas the spinwave line width is an indication of its power handling or nonlinear point. In practice, the resonance line width is used because of ease of accurate measurement, and, as can be seen, this is a rough measure of the other two properties except in the case of cobalt (Co), which can maintain an unchanged or very small increase in resonance line width but behaves like a rare earth relaxer otherwise. This has led to the false impression that it can reduce or leave the insertion loss unchanged while increasing power handling with increasing doping levels. In practice, it has about the same effect as samarium (Sm) or holmium when the ratio of increase in effective line width to spinwave line width is considered. Note that terbium (Tb) and dysprosium have about a 50% higher ratio and are thus less effective. Gadolinium is not considered a "fast" relaxer, but does gradually increase insertion loss and has a less favorable ratio than samarium or holmium. However, its temperature compensation properties are so useful this is often tolerated.

Both of these types of substitution are used in below-resonance devices almost exclusively. In above-resonance devices, fast relaxers are not required to improve peak power handling as spin waves are essentially suppressed and first-order nonlinear effects do not occur, although we will explore this assumption further in Chapter 8. Both fast relaxers and Gd increase the ferromagnetic line width substantially and make it difficult to operate near resonance, a major bandwidth limitation in above-resonance devices.

In below-resonance devices, it is important to emphasize that the above-mentioned levels of substitution will always proportionally raise the device insertion loss. For typical devices, every 10% of Gd for Y will raise the insertion loss by about 0.05 dB in a junction device under normal bias conditions. In high-power devices it is more difficult to generalize about the impact of Ho, but its effect is roughly 100 times greater than Gd.

It is also important to appreciate that Ho doping proportionally increases the peak power handling only, through its effect on spin wave line width. Its effect on mean power is complicated by heating effects. The best approach is to test the mean power at the doped and undoped levels, as most materials are available in both forms, to establish the correct balance. In critical devices, it may be necessary to adjust the doping level very precisely. Too little doping may cause nonlinearity below the desired power level; too much may cause high insertion loss and hence excessive heating. The fast relaxer effect is reduced at higher temperatures, and nonlinearity may again ensue as the device heats up, creating the potential for "runaway" catastrophic failure. In this case, heat sinking of the device and means of distributing the ferrite material over a larger area must be reconsidered.

In devices handling many KW of power, differential phase shift structures with greater surface must be used rather than junction devices, usually in waveguide. The effect of Ho on insertion loss and nonlinear threshold in this type of device can be seen schematically in Figure 1.5. Devices using high power are discussed in detail in Chapter 8.

Two nonlanthanides have been used on the dodecahedral site. Calcium (Ca) is commonly used to compensate for vanadium's (V) valency, and bismuth (Bi) has sometimes been used in single crystal and polycrystalline materials; combinations of these ions have limited levels of substitution.

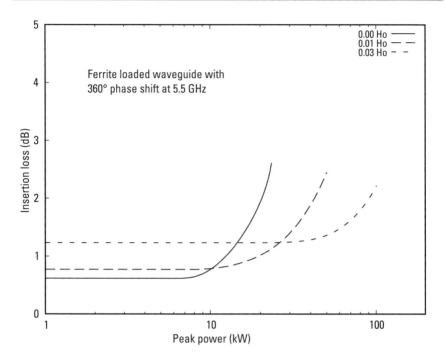

Figure 1.5 Effect of holmium doping on peak power handling and insertion loss in differential phase shifters.

1.6 Octahedral Substitution

In the late 1960s through the 1990s a series of newer substitutions on the octahedral site changed the world of above-resonance devices. Beginning with indium (In), it was found that the magnetocrystalline anisotropy of c-site Ca and a-site V substituted garnets could be reduced to essentially zero, creating polycrystalline materials with very low line widths, as low as a few Oersteds [5].

1.6.1 Nonmagnetic Octahedral Substitution

Indium was followed by tin (Sn), titanium (Ti), zirconium (Zr), and even antimony (Sb) as alternatives [5, 6]. Today, virtually all commercial materials use Zr on the octahedral site because of its lower cost compared with In. Sn is volatile and can be lost in manufacturing, Ti has limited octahedral substitution, leaking into the a-site, and Sb is toxic. In also has limited substitution but is used in the range 10–20% substitution for d-site Fe, well within its limits. The mechanism for all such ion substitution is the same; the reduced magnetization of the d-site lattice results in higher net magnetization of the garnet, and by changing the magnetocrystalline environment of the ferric ions also reduces the anisotropy and hence the ferrimagnetic line width of the material. Another unusual example is chromium (Cr), which enters the octahedral site up to 2.5%. Although it has three unpaired electrons it does not change the anisotropy or line width of YIG significantly.

These materials allowed wider bandwidth and lower losses in above-resonance devices from 50 MHz to 3 GHz. However, there were some drawbacks, mainly lower Curie temperatures for materials of the same 4πMs as YIG and Aluminium substituted YIG, reducing the potential temperature range and effective bandwidth (Figure 1.6).

In below-resonance devices particularly, despite the lower line widths, there is no evidence either from off resonance magnetic loss

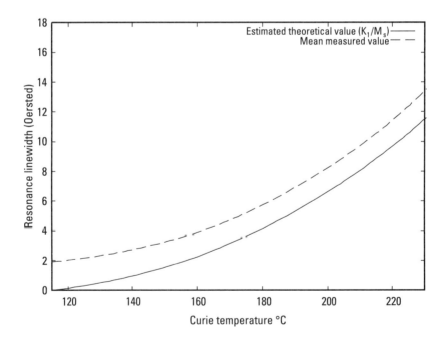

Figure 1.6 Effect of nonmagnetic octahedral substitution on resonance line width and Curie temperature for Zr, V substituted garnets, 4πMs 1,000—1,600 gauss.

measurements or practical devices that insertion losses are lower than, for example, YIG or Al substituted garnets under normal conditions. A further consideration in manufacturing garnets is the difficulty in controlling the dielectric losses of multivalent systems containing trivalent iron and yttrium but also tetravalent zirconium, divalent calcium, and pentavalent vanadium. Typical values are well below 0.0002, but by contrast YIG, which contains only trivalent ions, is routinely capable of much less than 0.0001. As a result of these factors, these narrow line width materials are not widely used below resonance. Nevertheless, because most communication devices are in the range 400 MHz to 3 GHz, and as a result are operated above resonance, a good proportion of infrastructure devices use Zr substituted V garnets with a range of $4\pi Ms$ of 800 to 1,950 gauss. Cellular radio handset devices use the lowest of these values to minimize magnet strength and size.

1.6.2 Manganese (Mn) Substitution

When Mn substitutes into the b-site it behaves as a Jahn-Teller ion, that is its bond lengths vary depending on its crystallographic environment, creating useful magnetoelastic effects like reducing magnetostriction in YIG and Gd Al garnets [7–9]. Jahn-Teller cations are a special case of octahedrally coordinated ions (see Figure 1.3) where the axial ratio can be stretched to accommodate changes in the lattice structure. Magnetostriction occurs when an external mechanical force is applied to magnetic materials, and their magnetic properties are changed by the resulting change in dimensions. The maximum change in magnetization at magnetic saturation is expressed as the magnetostriction coefficient, and is typically of the order of 10^{-6} of the linear dimensions.

Such magnetic and mechanical interactions are referred to as magneto-elastic properties. In structures under stress such as toroids in a waveguide where the garnet may be mechanically clamped in position, the saturation characteristics of the garnet are changed by the external pressure. In switching and latching devices this is a serious problem that is usually reduced by substituting around 4.5% of the b-site iron with Mn. Machining of toroids may result in similar stresses being set up in the surface that can only be removed by annealing even in Mn substituted materials [10]. Because Mn has almost the same magnetic moment as Fe (4.2 Bohr magnetons versus 5) and similar temperature behavior, the properties of Mn substituted garnets are almost the same as the unsubstituted types. Excess Mn can introduce higher dielectric loss because it can have valencies that include 2, 3, or 4, limiting the level of substitution.

It is now thought that magnetostriction may play a part in preventing saturation of garnets even in DC biased situations, with potential consequences for nonlinearity.

1.6.3   Cobalt (Co) Substitution

Cobalt enters the octahedral site in the divalent state, which can be induced by simultaneous equimolar substitution of tetravalent silicon (Si) or germanium (Ge). Without those trivalent Co will be created, with very different magnetic properties. This is important because only in the octahedral situation will it display strong positive anisotropy, in theory potentially canceling out the small positive anisotropy of Fe in YIG and Al or Ga substituted garnets. There is evidence that this occurs at small cobalt doping levels, but care should be taken in interpreting this as lower magnetic losses below resonance. As mentioned previously, it will improve the onset of nonlinear behavior with increasing power levels and is useful in this role (Figure 1.5 and Table 1.1) but will increase insertion loss because of the increase in spinwave line width.

1.7   Tetrahedral Substitution

Nonmagnetic ions on the tetrahedral site will reduce the overall magnetization by reducing the dominance of the site over the octahedral site, while also gradually decreasing the Curie temperature.

1.7.1   Aluminum (Al)

Initially, only Al was used in the tetrahedral site to reduce magnetization in YIG from its accepted value of 1,780 gauss. This works well up to about 35% of replacement for ferric ions on this site, or down to about 800 gauss; above that level the Al begins to leak into the octahedral site. The net effect is that progressively more Al is needed to make the net magnetization go down as both site sublattices add the nonmagnetic ion. The Curie temperature also falls, because of all of the nonmagnetic ions present, and this eventually prevails in reducing the overall magnetization (Figure 1.7).

This means that low magnetization Al garnets (less than 400 gauss) have very low Curie temperatures [11], making them difficult to use. They do, however, have good line widths (Figure 1.8) because the octahedral Al behaves like In or Zr in reducing anisotropy (K1). Fortunately, this corresponds to frequencies less than 2 GHz in below-resonance devices, so for most applications they are replaced by above-resonance devices. Exceptions are very broadband devices, typically close to an octave bandwidth, where below-resonance devices have to be used. The resultant devices perform very poorly over wide temperature ranges because of the loss of bandwidth at high temperature as the $4\pi Ms$ falls.

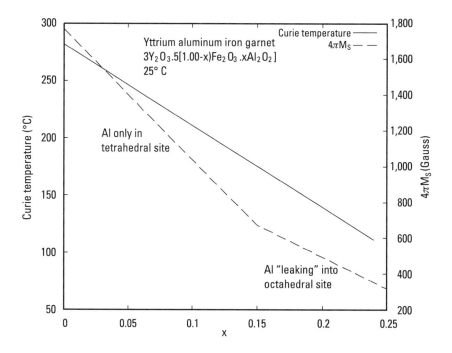

Figure 1.7 Aluminum substitution in YIG. (*After:* [11].)

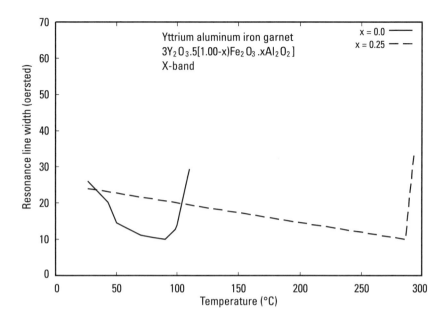

Figure 1.8   Resonance line width of YIG and aluminum-doped garnets with temperature. (*After:* [11].)

1.7.2   Gallium (Ga)

An alternative is Ga, which behaves better than Al in staying on the tetrahedral site at high substitution, and produces low magnetization material with reasonable Curie temperatures. Because no Ga enters the octahedral site, the anisotropy rises because K1/Ms is then determined by Ms and higher line widths result, reducing the material's usefulness. Cobalt can be used to reduce the resonance line width somewhat [12]. In practice because of the high cost of Ga, such materials are rarely used.

1.7.3   Vanadium (V)

As mentioned, V is used on the tetrahedral site in tandem with non-magnetic octahedral substitution, usually Zr, to produce low line width materials. Ca is used to balance up valencies and appears on the dodecahedral site. Part of the reason that Ca and V are used is that V does not reduce the Curie temperature as much as other nonmagnetic tetrahedral substitutions.

Most ranges of practical materials are made by fixing the Zr content and varying the V content to adjust $4\pi Ms$. This will work for $4\pi Ms$ in the 1,000 to 1,950 range, giving Curie temperatures around 200°C, and line widths less than 10 Oersteds (Figure 1.6). Clearly this will give a wide range of $4\pi Ms$ versus temperature slopes and limits the usefulness of some of these materials. Not surprisingly, the trend has been to use YIG itself instead because of its higher Curie temperature (280°C), with the emphasis on reducing YIG's line width by processing it to near theoretical density. Even with this approach, however, line widths for YIG of significantly less than 20 Oersteds are difficult to consistently achieve by conventional methods.

1.8 Mixed Systems

In general the tendency in industry has been to use either YGdAlFe garnets or YCaZrVFe garnets. In practice it is possible to mix these to give combinations of moderately lower line widths and temperature stability using Gd particularly [13, 14]. In practice this has not fitted either device situation. Above-resonance devices tend to need a low line width and linear $4\pi Ms$ versus temperature slopes, to allow the very strong biasing hexagonal ferrite magnet to track relative to resonance with the ferrite, not the flatter but much more nonlinear slopes associated with Gd substitution. Below-resonance devices in general do not need very low line widths but do need flat $4\pi Ms$ versus temperature characteristics, as in this case biasing fields allow weaker but more temperature-stable metal magnets.

One of the considerations in choosing below-resonance materials (Chapter 8) is the increase of line width with decreasing temperatures. This creates higher insertion loss on the low frequency side and will represent the limit of what can be used. This situation also arises in above-resonance devices but on the high-frequency side. Hence there are limits on Gd substitution below resonance because of the influence of the compensation point. The line width is effectively infinite at that point and that effect tails upwards towards room temperature. In this same way, the increase in line width of Al garnets with reducing temperature may limit their use, even if the room temperature value is acceptable.

1.8.1 Low Firing Temperature Garnets

Although bismuth and vanadium doped garnets have low firing temperatures [15], they have less useful properties than aluminum and gadolinium doped equivalents. The latter can have lower firing temperatures when fired with Cu [16, 17] and useful properties are obtained at temperatures of around 1,050°C, of interest for thick film–based low temperature cofired ceramic (LTCC) applications where ceramic dielectric and ferrites could be combined.

1.9   Rare Earth Substitution

In 2010, it became clear there would be a significant reduction in the supply of rare earth elements, which are processed mainly in China. As a result, the cost of rare earth materials has risen substantially, and supply is uncertain. Substitution of all or some of the yttrium in garnets is therefore of potential use commercially. As discussed earlier, there are at least two potential strategies to replace Y. This includes direct replacement with bismuth, which is trivalent, in the dodecahedral site. This has been done up to at least as far as $Bi_{0.71}Y_{2.29}Fe_5O_{12}$, but combinations of bismuth and vanadium have limited substitution limits, and single phase material is not obtained for y < 0.96 in Bi3-2yCa2yFe5-yVyO12, which has the effect of limiting the available magnetization with reasonable line width (<100 Oersteds) to 300 to 600 gauss. This cannot be overcome by using octahedral substitution to raise magnetization and reduce anisotropy and hence line width because the octahedral sublattice is already dominant over the tetrahedral due to the high V substitution. As a result, research is currently aimed at finding combinations of bismuth substitution, with low or no V tetrahedral content, and nonmagnetic octahedral substitution, as in this case the tetrahedral sublattice is dominant. Because such substitution allows the use of quadravalent or pentavalent ions on the octahedral site, potentially high magnetization is possible with low line width, along with substitution of divalent calcium for yttrium to further reduce the latter's content.

1.10 Summary

We have seen that a large variety of cation substitution is possible in the YIG structure, and we will examine how these are used in subsequent chapters. Tables of the most useful and well-characterized substitutions give a more concise picture (Tables 1.2 and 1.3). In today's world, only the Y, Gd, and Al series, with or without Ho or Mn, are used in high volume for below-resonance applications. Examples are radar and point-to-point microwave radios for cellular infrastructure. Above resonance, only the Al doped series and the Zr, V doped series have found extensive commercial application, mainly for cellular base station devices, where they are invariably used. Upwards of 100 million garnet- or spinel-based devices are manufactured each year for these applications. A similar number are used for cellphone handset applications, but only for some transceiver architectures.

Table 1.2
Iron and Nonmagnetic Substitution in Garnets

Metal Cation	Valency	c-Site (Dodec.)	a-Site (Oct.)	d-Site (Tetr.)	% Site Subst.	Comments
Iron (Fe)	2,3		X (+3)	X (+3)	100	+3 only desirable
Y	3	X			100	Nonmagnetic
Ca	2	X			100	+4,+5 Valency Comp.
Al	3		X	X	100	Leaks d to a above 35% d
Ga	3		X	X	100	Fills d-site first
Ge	4			X	>15%	Expensive
V	5			X	>50%	Favorable Curie temperature
In	3		X		<40%	Reduces LW with V, Ge
Cr	3		X		~3%	Limited use
Bi	3	X			100%	Limited combinations
Zr	4		X		>35%	Reduces LW(V), cheap
Sn	4		X			Reduces LW, volatile
Ti	4		X	X		Leaks a to d, 7:3 ratio
Sb	5		X		−75%	Experimental, toxic

Table 1.3
Magnetic Substitution in Garnets

Metal Cation	Valency	c-Site (Dodec.)	a-Site (Oct.)	d-Site (Tetr.)	% Site Subst.	Comments
Mn	2,3,4		X		>7%	+3 is Jahn-Teller ion
Co	2,3		X (+2)	?		Fast relaxer (+2)
Gd	3	X			100%	Compensation point, weak relaxer
Ho, Dy, Tb	3	X			100%	Compensation point, fast relaxer
Sm	3	X			100%	L-S coupled, fast relaxer

References

[1] Geller, S., "Crystal Chemistry of the Garnets," *Zeitschrift fur Kristallographie,* Vol. 125, 1967, pp. 1–47.

[2] Bertaut, F., and F. Forat, "Structure des Ferrites Ferrimagnetiques des Terres Rares," *C. R. Acad. Sci.,* Vol. 242, 1956, p. 382.

[3] Goodenough, J. B., *Magnetism and the Chemical Bond,* New York: Interscience-Wiley, 1973.

[4] Inglebert, R. L., and J. Nicolas, "Relaxation of Rare Earth and Cobalt Ions in Polycrystalline Yttrium Iron Garnet," *IEEE Trans. Mag.,* Vol. MAG-10, No. 3, 1974, p. 610.

[5] Van Hook, et al., "Linewidth Reduction Through Indium Substitution in Calcium-Vanadium Garnets," *J. Appl. Phys.,* Vol. 39, No. 2, 1968, p. 730.

[6] Winkler, G., "Substituted Polycrystalline YIG with Very-Low Ferrimagnetic Resonance Linewidth and Optical Transparency," *IEEE Trans. Mag.,* Vol. MAG-7, No. 3, 1971, p. 773.

[7] Cruickshank, D. B., "Antimony Substitution in Garnets," *Physical Society Colloquium on Magnetic Oxides,* Imperial College, London, 1970.

[8] Dionne, G., "Effect of External Stress on Remanence Ratios and Anisotropy Fields of Magnetic Materials," *IEEE Trans. Mag.,* Vol. MAG-5, No. 3, 1969, p. 596.

[9] West, R. G., R. L. Huntt, and A. C. Blankenship, "Control of Magnetostriction Effects in Polycrystalline YIG," *IEEE Trans. Mag.,* Vol. MAG-4, No. 3, 1968, p. 610.

[10] TransTech Tech Brief No. 691, "Stabilization of Remanent Induction by Thermal Annealing."

[11] Harrison, G. R., and L. R. Hodges, "Microwave Properties of Polycrystalline Hybrid Garnets," *J. Amer. Ceram. Soc.*, Vol. 44, 1961, p. 214.

[12] Cruickshank, D. B., "Current State of the Ferrite Materials Art," *IEEE MTTS*, Vol. WML-02, 2008.

[13] Hudson, A. S., "Substitution of Gadolinium, Aluminum and Indium in Yttrium Calcium Vanadium Iron Garnets," *Intermag. Conf. Abs.*, Vol. 28.1, 1969.

[14] Nicolas, J., and A. LaGrange, "Magnetic and Microwave Properties of Polycrystalline Yttrium-Calcium-Gadolinium-Iron-Tin Garnets," *Proc. Int. Conf. Ferrites (ICF)*, 1970, p. 527.

[15] Wilson, W. R., L. R. Hodges, and G. P. Rodrigue, "Magnetic and Microwave Properties of Aluminum- and Gadolinium-Substituted Calcium Vanadium Garnets," *J. Applied Phys.*, Vol. 38, No. 3, 1967, p. 1405.

[16] Cruickshank, D. B., "Ferrites for Switching Applications," *IEE Colloquium on Ferrite Materials, Devices and Applications*, Edinburgh, Vol. 3, 1989, pp. 1–5.

[17] Lebourgeois, R., et al., "Low Sintering Temperature Garnets for Microwave Integrated Circuits," *IEEE MTTS*, Vol. WSF-11, 2005.

Selected Bibliography

Inui, T., and N. Ogasawara, "Grain Size Effects on Microwave Ferrites Magnetic Properties," *IEEE Trans. Mag.*, Vol. MAG-13, No. 6, 1977, pp. 1729–1744.

Von Aulock, W. H., *Handbook of Microwave Ferrite Materials*, New York: Academic Press, 1965.

Winkler, G., *Magnetic Garnets*, Braunschweig Vieweg, Germany, 1981.

2

Spinels

2.1 Introduction

The spinel family takes its name from the mineral spinel ($MgAl_2O_4$), which occurs abundantly in nature (Figure 2.1). The space structure was discovered by Bragg, and separately by Nishikawa, in 1915. The unit cell in magnetic spinels contains eight molecules of $MeFe_2O_4$ where Me is typically a divalent transition metal ion. However, it was progressively discovered that that these magnetic spinels fell into three categories of structure, namely normal, inverse, and a mixture of the two.

To understand the differences between these categories, part of the unit cell is shown in Figure 2.2, showing the two types of metal ion coordination, octahedral and tetrahedral. The complete unit cell consists of eight octants of Figure 2.2, in alternating positions. However, of the available 64 octahedral and 32 tetrahedral sites, only 8 and 16, respectively, are filled to form the unit cell, that is $Me_8Fe_{16}O_{32}$ if "normal," as we will see in the following sections of this chapter.

In the mineral spinel ($MgAl_2O_4$), all of the occupied tetrahedral or A-site ions are divalent magnesium ions, and all of the occupied octahedral or B-site ions are trivalent aluminium. This can be represented as $Mg^{+2} [Al_2^{+3}] O_4$, where [ ] is the octahedral position. This structure became what is called a "normal" spinel.

Figure 2.1   Naturally occurring spinel crystal.

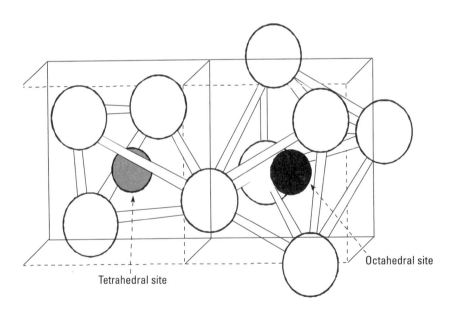

Tetrahedral site

Octahedral site

Figure 2.2   Part of the spinel structure.

As we know from Chapter 1 on garnets, not all Fe^{+3}, for example, will go into the smaller tetrahedral sites, and similarly in garnets, Al^{+3} will progressively leak from tetrahedral to octahedral sites as the Al^{+3} content increases. In the case of the spinels with Fe^{+3}, normal, inverse, and mixed occupation by Fe^{+3} can be represented as:

Normal: $Me^{+2}[Fe_2^{+3}]O_4$
Inverse: $Fe^{+3}[Me^{+2}Fe^{+3}]O_4$
Mixed or partially inverse: $Me_{(x)}^{+2}Fe_{(1-x)}^{+3}[Me_{(1-x)}^{+2}Fe_{(1+x)}^{+3}]O_4$

This distinction, verified by neutron diffraction, is important to understanding the magnetic properties of spinels. In general, smaller ions prefer A-sites and larger ions prefer B-sites, but the electron configuration of Zn^{+2}, for example, favors tetrahedral sites, even though Zn^{+2} is relatively large. Electrostatic energy factors, including the oxygen u factor, also influence site preference [1].

From the point of view of RF applications, we will only consider a fraction of the possible spinel compositions. These include nickel, magnesium, and lithium spinels (where divalent Me is represented by ($Li^{+1}Fe^{+3}$), and those three major microwave and RF types of spinels, which also contain varying substitutions of copper, cobalt, zinc, manganese, aluminum, and titanium. Of those substitutions, nickel, cobalt, and copper are essentially inverse, although the latter two are used only at relatively low doping levels. Magnesium is partially inverse, about 90% or $x = 0.1$ in the mixed case above, while manganese is about 20% inverse, or $x = 0.8$. Lithium ferrite spinel is completely inverse as $Fe^{+3}[Li_{0.5}^{+1}Fe_{1.5}^{+3}]O_4$.

2.2 Nickel Spinels

Predicting the magnetization is relatively straightforward for most ferrimagnetic spinels. For nickel spinels, where Ni^{+2} has eight electrons in its d-orbital, there are two resultant unpaired electrons. Since nickel spinel is inverse, the overall moment is determined by the difference between one octahedral B-site Fe^{+3} and the sum of one tetrahedral A-site Fe^{+3} and one A-site Ni^{+2}. The resultant theoretical moment of two

Bohr magnetons at low temperature assumes this simple A-B interaction, and measured values indicate it is essentially correct. In practice, more complex A-A and B-B exist, and it was found [2] that a few percent of Ni^{+3} ions are on B-sites, modifying the behavior of these ions such that they behave as "fast relaxers" (see Section 1.5).

We can increase the magnetization of nickel spinels by substituting nonmagnetic ions like zinc for iron on the A-site, and reducing the number of Ni^{+2} ions on the B-site. In this case the Zn^{+2} ion does not completely show its preference for A-sites as theory would predict. The reduced A-site magnetization increases the net molecular moment as the net B-site Ni^{+2} and Fe^{+3} further dominate. Eventually, as the Curie temperature falls with increasing zinc, the room temperature magnetization falls again, producing a maximum magnetization of about 5,000 gauss when around 40% of Ni^{+2} is substituted by Zn^{+2}. This also produces a reduction in magnetocrystalline anisotropy (K1/Ms) analogous to the effect of nonmagnetic ions on the octahedral site in garnets, albeit on the tetrahedral site in the spinel. It is likely also that Zn^{+2} reduces the small percent of Ni^{+2} on the B-sites, as intrinsic magnetic losses in the NiZn spinel would seem to indicate a reduction in the "fast relaxer" concentration. Small amounts of In^{+3} or Ga^{+3} have the same effect as Zn^{+2} by increasing magnetization [3, 4].

Conversely, the magnetization of spinels can be reduced by nonmagnetic substitution of octahedral magnetic ions. Hence Al^{+3} in Ni and Mg spinels and Ti^{+4} in Li spinels have this effect. This gives a useful range of magnetizations of about 600 to 5,000 gauss for spinels, compared to a range of 300 to 1,950 gauss for garnets. The main limiting criterion is the Curie temperature, which ideally should be well above 100°C for the low end of the magnetization range, above 200°C for intermediate values, and over 300°C for the high end of the range. As a result spinels are not used much below 2,000 gauss magnetization today.

2.2.1 Overview of Nickel Spinel Applications

Although in the past Ni spinels have covered the entire range of RF frequencies, in practice today this is largely confined to two separate ranges, from 10 MHz (the lowest frequency we will consider for this publication) to about 200 MHz, and an upper range from about 10

GHz to well into the millimetric range. The exception is largely related to ferrite absorbers (Chapter 3) where domain and spin resonance phenomena allow relaxation absorption in unmagnetized, or in some cases partially magnetized, ferrite. The lower limit for absorbers is broadly determined by Snoek's limit, determined in principle by permeability and magnetization, which we will discuss elsewhere, and the upper limit by "low field loss," which is determined by magnetization, the gyromagnetic ratio, and shape-related demagnetization factors. As a result, NiZn-based absorbers are used from about 10 MHz to 12 GHz.

At low frequencies, pure Nickel ferrite has good permeability, approximately 30, and low magnetic loss, to more than 50 MHz, and can be extended in frequency by reduced firing temperatures or time to beyond 100 MHz because of the effect of porosity and fine grains, albeit with reduced permeability. This effect was first noted by Schettler [5] who found that firing temperature (the equivalent of density and grain size), shrinkage (also affecting density and grain size), and Zn^{+2} and Co^{+2} additions all affected the frequency of the first relaxation peak. He attempted to analyze these contributions mathematically but the interdependence of density and grain size obscured the results.

At about the same time Pippin [6] looked at a similar range of compositions. He suggested that Snoek's formula only covered at best the NiZn ferrites, where Zn reduced both magnetocrystalline anisotropy and magnetostriction, such that the increased permeability and magnetization due to Zn^{+2} could be inversely related to the first domain wall resonance frequency. When Co^{+2} is introduced, the first-order magnetocrystalline anisotropy constant (K1) reaches a minimum value close to 0 at 2.7% Co^{+2} for Ni^{+2}, which Pippen suggested would give the maximum frequency shift and highest permeability if K1 were the only contributor. In practice, Pippen suggested, the anisotropy constant K2 is not 0 at 2.7% Co^{+2} and magnetostriction is increasing rapidly more negatively with increasing Co^{+2} substitution, offsetting the K1 contribution. Reproducing these results with modern methods of stoichiometry control and firing conditions indicate that there is monotonic reduction of permeability with Co^{+2} additions initially, followed by a leveling off or slight increase where K1 would be expected to be close to 0, then followed by an increasing rate of drop in the range 2.5 to 4% and beyond of Co^{+2}. The frequency shift follows the same pattern. It is also possible to raise the first resonance frequency by decreasing density and grain size, but this must be held to practical controllable limits to prevent open porosity and hence moisture absorption. Of the three contributing mechanisms, it can be said that magnetocrystalline anisotropy is the best characterized. Ni ferrite is slightly negative, and can be balanced by small amounts of strongly positive Co^{+2}; this amount will be reduced in NiZn ferrites because nonmagnetic Zn^{+2} weakens the negative anisotropy progressively as its concentration increases. The rotational susceptibility is proportional to $Ms^2/K1$ for a dense, single phase ferrite, and hence the domain rotation frequency shows the same behavior. Since Co^{+2} has essentially no effect on the magnetization, the rotation frequency should show a maximum when K1 approaches zero at the compensation point of Co^{+2} for each Ni/Zn ratio. This effect is seen at the first and second frequencies in Figure 2.3 but less so at F1, the lowest frequency, for a 5,000 gauss Ni/Zn ferrite with increasing Co^{+2} doping. The most striking effect is on the permeability, which plateaus when the magnetocrystalline anisotropy is close to 0 for Cobalt doping around 0.015 to 0.025, before falling at higher doping levels from 50 to below 10 (not shown). Table 2.1 also shows Pippin's original data on Co^{+2} doped Ni ferrite in the same form, where similar behavior is observed.

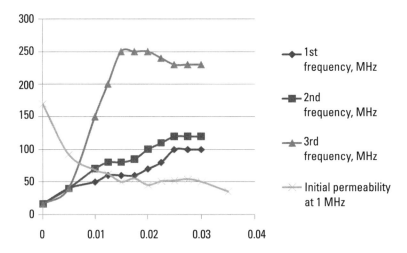

Figure 2.3   Initial permeability and first, second, and third resonance peaks (Y-axis, MHz) of Co doped (X-axis) NiZn spinel.

Table 2.1

Effect of Cobalt substitution in $Ni_{1-x}Co_xMn_{0.02}Fe_{1.9}O_4$ Spinel on Spectra

Cobalt Substitution	First Frequency (MHz)	Second Frequency (MHz)	Porosity %	Initial Permeability at 5 MHz
0.00	100	1,700	4.5	28
0.01	270	1,400	5.0	17
0.027	280 (merged)	merged	4.8	20
0.036	500 (merged)	merged	5.5	16
0.068	550	2,500	10.2	8.5

Source: [6].

Magnetostriction, which is negative in Ni ferrite, is also weakened while remaining negative in proportion to Zn^{+2} content, is increased further by the negative contribution from Co^{+2} doping. This should effect the domain movement monotonically as Co^{+2} concentration increases, and may be the reason the first frequency in Figure 2.1 continues to rise as Co^{+2} increases, although with a flat region in the K1 ~ 0 range. This is in line by the assertions of Pippin [6], Rao [3], and later Mikami [7] that the first, lowest frequency in the magnetic spectrum is dominated by domain wall movement, not rotation. For many applications, it is therefore important to adjust the cobalt doping to the optimum value of the first frequency appropriate to the Ni/Zn ratio, such that the highest permeability is obtained for the highest first frequency possible, and hence lowest magnetic loss.

Reducing the density of Ni and NiZn ferrites also increases the frequency of the first relaxation peak frequency, but with a corresponding drop in initial permeability. Pippin's results are summarized in Tables 2.2 and 2.3, with some frequency data added showing the effect on a combination of lowered density and Co^{+2} doping. Reduced density has the advantage of retaining low magnetostriction and temperature dependence of K1 compared to Co^{+2} doped equivalents. Because reduced density is normally associated with smaller grains in the firing process, older data typically cannot distinguish between the two effects. However, Igarashi and Okazaki [8] used hot pressing to achieve dense fine-grained specimens and were able to assess the

Table 2.2
Effect of Porosity on Spectra of $NiMn_{0.02}Fe_{1.9}O_4$ Spinel on Spectra

Porosity %	First Frequency (MHz)	Second Frequency (MHz)	Initial Permeability at 5 MHz
4.6	100	1,700	28
7.7	170	1,600	20.5
9.8	330 (merged)	Merged	17
14.7	350 (merged)	Merged	13
29.2	650 (merged)	Merged	6.5

Source: [6].

Table 2.3
Effect of Porosity on Spectra of $Ni_{(0.96)}Co_{0.04}Fe_{1.94}O_4+/-$ Spinel

Porosity %	First Frequency (MHz)	Second Frequency (MHz)	Third Frequency (MHz)	Initial Permeability at 1 MHz
4.85	130	~400	1,400	15.3
6.8	300	800	1,600	8.8

Source: [6].

relative contribution. They showed that decreased permeability was a function of porosity and an inverse function of the demagnetization factor. They used a modified version of the Globus expression to derive a relationship between observed decrease in permeability and the grain diameter (radius), magnetization, and volume change of the magnetic domain, giving a dependency of permeability proportional to the cube root of the grain size. The change of first relaxation frequency (Fr) with porosity and grain size could then be calculated to first order using the measured initial permeability μ, since most expressions reduce to

$$Fr = Constant \times 1/\mu - 1 \qquad (2.1)$$

for a fixed composition with essentially fixed K1 and magnetization, when it is within 10% of its theoretical density. However, we can assume theoretically for a dense ferrite that only the domain wall movement contributes, and for very porous ferrite rotation contributes or dominates, although some modification of these assumptions might apply at intermediate densities. See [7] for a more general discussion.

2.2.2 Nickel Ferrites Above 10 GHz

Nickel ferrites are used below resonance from about 10 GHz to approaching 100 GHz. Using the relationship between the lowest frequency and gamma x $4\pi Ms$, this is equivalent to the range of 3,000 to 5,000 gauss, the latter being the highest possible magnetization from the NiZn system. In practice pure Ni ferrite (3,000 gauss) is not used much for junction devices because of its higher insertion loss, which can be attributed to A-site Ni^{+2}, unless a higher spinwave linewidth is required for power handling; Mg ferrites are used instead. The NiZn series, however, does have low losses because of reduced anisotropy and is used from 3,500 gauss upward. Both Ni and NiZn are used for high-power differential phase shifters because these devices typically handle high power. Ni ferrites have quite high magnetostriction and poor BH loop squareness. NiZn ferrites have progressively lower magnetostriction and at higher Zn levels can have square loops and low magnetostriction through the addition of manganese, and can be used for switching or latching applications (Chapters 8 and 11). The remanence/saturation ratio of the hysteresis loop can be improved from about 0.6 to > 0.9 in this way.

NiAl ferrites have lower magnetization and lower losses than Ni ferrite, presumably through the reduction of A-site Ni^{+2}. The gyromagnetic ratio rises with increasing Al substitution, reaching a maximum value at about 0.62 Al for Fe in the Ni spinel formula, an effect used successfully with resonance isolators to reduce the applied external field. However, most current applications requiring lower magnetizations use garnet compositions because garnets have lower dielectric losses, and more controllable magnetic losses, with the option to use holmium to raise the spin wave linewidth. Both the spinwave linewidth and the gyromagnetic ratio of Ni and NiAl ferrites have been increased in the past by using underfired material with fine grains and substantial porosity. These types of materials are less strong mechanically and may absorb water, so this practice has largely been discontinued, particularly for high-power applications where these properties are extremely undesirable.

Note that Co^{+2} can be used to raise the power handling of nickel ferrites, acting as a fast relaxer in both spinels and garnets (see Chapter 1). It is used in the same way in magnesium and lithium ferrites [9]. Mn^{+2} is used in all spinels to reduce magnetostriction although not a Jahn-Teller ion (note that octahedral Mn^{+3} is the Jan-Teller used in garnets). Dielectric loss can also be reduced, thought to be by sacrificial reduction of Mn^{+3}, because of the possible $Fe^{+2} + Mn^{+3} = Fe^{+3} + Mn^{+2}$ reaction during sintering. Cu^{+2} is used as a sintering aid and enters the B-site. Cu^{+1} has been used to produce high $4\pi Ms$ materials, but the ease of oxidation of Cu to +2 makes this option extremely difficult to manufacture.

2.3 Magnesium Spinels

Magnesium spinel can give useful magnetizations in the range 700 to 3,000 gauss using Al to lower and Zn to raise magnetization over its unsubstituted value of about 2,200 gauss. Manganese doping is used to minimize dielectric losses and improve the BH loop. These ferrites have very low magnetic losses away from resonance (comparable with garnets), but their broad line width and variable magnetization from processing has limited current use to a few applications below resonance where their low losses and useful magnetizations (2 to 3,000 gauss) or square loop make them the best choice. The magnetization variability relates to the dependence of the degree of inversion upon the fired ceramic body cooling rate. The higher line width is due to a number of factors: ordering or clustering of magnesium ions, being second phase (usually magnesium oxide as periclase) rich, and difficultly reaching close to theoretical density. None of these factors, however, have much effect on magnetic losses away from resonance, and magnesium ferrites in junction devices give similar losses to YIG and Al doped YIG. A more quantitative study of losses was carried out by Green et al. [10, 11] by measuring magnetic losses of Mg spinel and YIG in a waveguide using varying biasing conditions. They concluded that a single value of magnetic loss, μ, could be used for all magnetization states below resonance provided the 4πMs value chosen avoided low field loss.

The microstructural aspects of many Mg ferrites tend to give improved power handling through spin wave scattering of pores, nonmagnetic second phase material, and finer grains, but can be variable because of different processing conditions, and for high-power applications Co^{+2} doping is necessary to ensure reproducibility of the spin-wave line width and hence peak power measurement.

Commercial magnesium ferrites are available in a range of magnetizations using Al to reduce and Zn to increase the value. Both reduce the Curie temperature (as would be expected) and the line width, so the basic magnesium ferrite is highest in Curie temperature. The Zn line width reduction is caused by reduced anisotropy and the Al reduction, by changing the degree of inversion and improving disordering (see Figure 2.4).

2.4 Lithium Ferrite

Pure lithium ferrite has been grown as single crystals and has low linewidth in that form, with a high magnetization (3,500 gauss) and Curie temperature (>600°C), making them virtually unique among the spinels and complementing single crystal garnets. Manufacturing of single crystals requires due attention to the loss of volatile lithium and

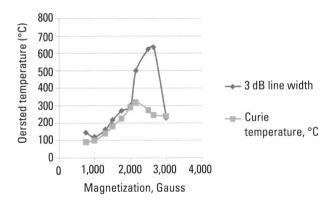

Figure 2.4   3-dB line width and Curie temperature (Y-axis) against magnetization in gauss for Al- and Zn-substituted magnesium ferrites.

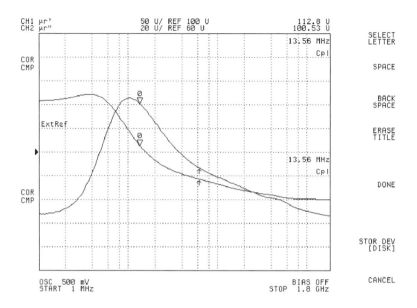

Figure 2.5   Magnetic absorption spectrum of a NiLiZn ferrite.

requires ordering by slow cooling through the ordering temperature of about 750°C. Ordering in this context means the correct positioning

Table 2.4
Magnetic Substitution in Spinels

Metal Cation	Valency	A-site (Tetr.)	B-site (Oct.)	% A-site	Comments Microwave	Comments VHF/UHF
Iron (Fe)	2, 3	X(+3)	X(+3) X(+2)	Up to 100%	+3 only desirable	+2 with +3 useful for e-based absorber
Mn	2, 3	X (+2)	X(+2)	80% (usually dopant)	Improves dielectric loss/ magnetostriction as dopant	Improves dielectric loss; Mn ferrite useful only < 1 MHz
Co	2		X	100% (usually dopant)	Improves peak power as dopant	Extends μ/μ frequency as dopant
Ni	2	X	X	Very small	High Curie temperature	Use μ with Zn

of the cations in their preferred sites. Polycrystalline lithium ferrite was first made manufacturable by the use of small amounts of bismuth oxide [9, 12], which reduced the firing temperature and allowed a controlled microstructure, preventing excessive loss of lithium. A wide range of magnetizations was achieved by using combinations of Ti to reduce magnetization and Zn to achieve good line width. Originally it was considered a rival to garnet for magnetizations in the 600 to 2,000 gauss for switching or latching applications on cost grounds, but rapid reduction in rare earth prices largely eliminated that argument, although lower magnetostriction is still a valid consideration. Its prime advantage now is for switching or latching applications at high values of $4\pi Ms$ and hence frequencies, where lithium ferrites can have

Table 2.5
Nonmagnetic Substitution in Spinels

Metal Cation	Valency	A-site (Tetr.)	B-site (Oct.)	% A-site	Comments Microwave	Comments VHF/ UHF
Li	1	—	X	—	High Curie temperature	Useful μ/e
Mg	2	X	X	Variable	Low loss	Not useful
Cu	1,2	—	X(+2)	—	Improves density as dopant	Influences microstructure/μ
Al	3	—	X	—	Lowers $4\pi Ms$	—
Bi	3	—	—	—	Improves density; Intergranular only	—
Cr	3	—	X	—	Mn ferrite study	—
In	3	X	—	—	—	As Zn
Zn	2	X	—	—	Raises $4\pi Ms$	Raises initial μ
Ti	4	—	X	—	As Al but improved switching	—
Zr	4	—	X	—	—	MnZn study
Sn	4	—	—	—	—	MnZn study
V	5	—	X	—	—	MnZn study
Nb	5	X	X	20%?	Li ferrite BH improvement	Cu^{+1} ferrite study; NiZn μQ study

higher Curie temperatures than the corresponding NiZnMn ferrites at compositions giving magnetizations around 5,000 gauss. On balance, however, the more complex manufacturing processes for lithium ferrites, particularly firing, make them less attractive.

Lithium ferrites have been combined with NiZn ferrites to make broadband absorbers, using the relatively easy reduction of Fe^{+3} to Fe^{+2} in an Li ferrite to create material with high dielectric and magnetic loss. Its magnetic spectrum (permeability μr and magnetic loss μr) is shown in Figure 2.5, measured on an impedance analyzer (Chapter 12).

2.5 Summary

A summary of the affects of cation substitution in spinels appears in Table 2.4 and 2.5.

References

[1] Venway, E. J. W., F. deBoer, and J. H. Van Santen, "Cation Arrangement in Spinels," *J. Chem. Phys.*, Vol. 16, 1948, p. 1091.

[2] Robertson, J., and A. J. Pointon, "Ni^{+2} Ion Distribution in Nickel Ferrites," *Solid State Communications*, Vol. 4, No. 6, 1966, p. 257.

[3] Rao, K. H., et al., "Effect of Indium Substitution on the Properties of Ni-Zn and Ni-Zn-Ti Ferrites," *ICF 9*, 2004, p. 365.

[4] Maxwell, L. R., and S. J. Pickart, "Magnetization in Nickel Ferrite Aluminates," *Physics Rev.*, Vol. 92, 1953, p. 1120.

[5] Schnettler, F. J., and F. R. Monforte, "Effect of Cobalt on the Relaxation Frequency of Nickel-Zinc Ferrite," *J. Appl. Phys.*, Vol. 29, No. 3, 1958, p. 477.

[6] Pippin, J. E., and C. L. Hogan, "Resonance Measurements on Nickel-Cobalt Ferrites as a Function of Temperature and on Nickel Ferrite-Aluminates," *IRE Trans. MTT,* Vol. 6, 1958, p. 1120.

[7] Mikami, I., "Role of Induced Anisotropy in Magnetic Spectra of Cobalt-Substituted Nickel-Zinc Ferrites," *Jap. J. Appl. Phys.*, Vol. 12, No. 5, 1973, p. 678.

[8] Igarashi, H., and K. Okazi, "Effects of Porosity and Grain Size on the Magnetic Properties of NiZn Ferrite," *J. Amer. Ceram. Soc.*, Vol. 60, No. 1-2, 1977, pp. 51–54.

[9] Baba, P. D., et al., "Fabrication and Properties of Microwave Lithium Ferrites," *IEEE Trans. Mag.*, Vol. MAG-8, 1972, p. 83.

[10]   Green, J. J., and F. Sandy, "Microwave Characterization of Partially Magnetized Ferrites," *IEEE Trans. MTT,* Vol. 22, No. 6, 1974, p. 641.

[11]   Green, J. J., and F. Sandy, "A Catalog of Low Power Loss Parameters and High Power Thresholds for Partially Magnetized Ferrites," *IEEE Trans. MTT,* Vol.-22, No. 6, 1974, p. 645.

[12]   Argentina, G. M., and P. D. Baba, "Microwave Lithium Ferrites: An Overview," *IEEE MTT,* Vol. MTT-22, No. 6, 1974, p. 652.

Selected Bibliography

Goldman, A., *Handbook of Modern Ferromagnetic Materials*, Boston, MA: Kluwer, 2002.

International Conference on Ferrites (ICF)—A series of 10 conferences from 1970 (ICF 1 Tokyo) to 2008 (ICF10 Chengdu) published individually. Latest published is *ICF 9 Proceedings*, American Ceram. Soc., 2005.

Smit, J., and H. P. J. Wijn, *Ferrites*, New York: John Wiley, 1959.

Van Groenou, B., "Magnetism, Microstructure and Crystal Chemistry of Spinel Ferrites," *Review Paper, Mater. Sci. Eng.*, Vol. 9, 1968, pp. 317–392.

Von Aulock, W. H., *Handbook of Microwave Ferrite Materials*, New York: Academic Press, 1965.

3

Absorbers

3.1 Introduction

The purpose of this chapter is to give a basis of comparison for various very different types of microwave absorbers, and to indicate the best applications for each one. Absorbers have been used for many years as microwave loads or to absorb stray or directed radiation in closed or open environments. The radiation required to be absorbed may be very broad spectrum or highly frequency selective. Bulk absorbers for microwave and VHF/UHF applications fall broadly into three types. The first is based on the relaxation phenomenon relating to dielectric or magnetic properties, the second to conductivity effects. A third type of absorber is based on powder and flake magnetic metals in sheet form and will not be discussed in detail for applications below 10 MHz, where they are often used. However, they have found applications for high-frequency noise suppression and are likely to become increasingly useful.

The prime applications for absorbers are for anechoic chambers, waveguide loads, absorbing sheets in microwave device and subsystem enclosures to absorb unwanted radiation, and radiation leakage around antennas. A separate topic is stealth, where broadband absorbing materials and coatings are used, but this will not be discussed in detail here. More recently both microwave radiation from devices and emitted

39

noise from very high-speed digital circuits are found within the same enclosure and the same absorber is required to handle both, often in very limited space such that relatively thin absorber sheets or coatings are needed.

We can compare absorber types using Table 3.1. Each broad type of material can be characterized by its absorption mechanism and the lowest and highest range of frequencies that the material can be affective. If magnetic, its initial permeability is given and the highest frequency that permeability is retained before the relaxation mechanism causes it to fall. If dielectric, the dielectric constant is given for either all frequencies if it does not change, or if it does, the dielectric constant at a frequency lower than that point and a second value at the application frequency. Its dielectric loss is given at its peak value within the application range of frequencies, which are also given in Table 3.1. Some materials have both magnetic and dielectric losses, often at different frequency ranges, so these are discussed separately. One issue with comparing absorbers is the basis for comparison. The dielectric and magnetic loss tangents will be used throughout this chapter as an absolute means of comparison, but many absorber materials are defined using parameters relating to their application. In addition, the dielectric constant and permeability do affect the reflected signal from the materials surface, and unless they are roughly equal or the part is electrically suitably impedance matched, the absorption will be substantially different from what might be expected solely on the basis of magnetic or dielectric loss. Absorbers are often backed with metal as part of the shielding process and that will often improve absorption because it can effectively double the material's absorption thickness because of reflection. Very sophisticated absorbers are also possible with successive layers of absorbers of different dielectric or magnetic properties.

3.2 Ni and NiZn Ferrite Absorbers

These ferrites were discussed in detail in Chapter 2. The principle absorption mechanism is magnetic loss, μ'' above the so-called Snoek frequency limit. In practice, a number of mechanisms operate at these frequencies. These mechanisms have been discussed in detail by a number of authors [1, 2] relatively recently. By using increasing amounts of Zn

Table 3.1
Types of Absorbers

Material	Mechanism	Absorption Range (Low)	Absorption Range (High)	μ Initial/Maximum Frequency	Dielectric Constant/Frequency	Dielectric Constant/Application Frequency	Maximum Dielectric Loss Tangent
Ni ferrite	Magnetic μ	200 MHz	10 GHz	30/200 MHz	13/all frequencies	13/10 GHz	<0.001
NiZn ferrite	Magnetic μ	1 MHz	15 GHz	4,000/1 MHz	13/all frequencies	13/10 GHz	<0.001
Water	Dielectric	<300 MHz	>100 GHz	1	87/300 MHz	32/100 GHz	>0.5/10 GHz
BaTiO$_3$	Ferroelectric	>10M Hz	~10 GHz	1	2,000+/<10 MHz	280/5.6 GHz	0.6/10 GHz
SiC	Semiconductor	<500 MHz	>20 GHz	1	>30/500 MHz	>30/20 GHz	0.5/20 GHz
Magnetic metal/polymer	Magnetic/dielectric/other	<1 MHz	>10 GHz	2 to 200/1 MHz	2 to 20+/1 MHz	2 to 20/10 GHz	Variable, >0.3/10 GHz
Hexagonal ferrite	Magnetic μ	500 MHz	Selective	5 to 20/up to 400 to 200 MHz	15/all frequencies	15/10 GHz	<0.01

it is possible to increase the initial permeability to in excess of 4,000 at 1 MHz, with the resulting relaxation-based drop in permeability and absorption peak at slightly higher frequencies. Reducing the amount of Zn will progressively reduce the permeability to about 30 at 0 Zn, while raising the absorption peak above 100 MHz. Hence we can tailor the absorption peak depending on the application. At frequencies above the first absorption peak, a series of other peaks are observed, which relate to spin or anisotropy-based resonances. These effectively extend the frequency range of the ferrite, as an absorber, to a value expressed by the product of the gyromagnetic ratio times the 4πMs, which is about 8 GHz for a 3,000 gauss ferrite and 12 GHz for a 5,000 gauss ferrite. As discussed previously, it was thought that Snoek's limit applied to NiZn ferrites fairly well, but the introduction of Co^{+2} with strong positive anisotropy allows us to manipulate the absorption peak with frequency without changing the 4πMs, although the permeability falls somewhat. This opens up the possibility of pushing the absorption peak higher, without overly compromising the permeability, at frequencies well beyond 200 MHz. We can also push the frequency up by reducing the density or grain size of any given composition [3–6], although the permeability will fall accordingly, and eventually this approach will produce an unacceptable density if open pores are produced and moisture ingress can occur. Typical NiZn ferrite absorption behavior is shown in Figure 3.1, in this case with the first absorption peak at about 13 MHz. This ferrite can also be modified to raise its dielectric loss by creating Fe^{+2} in the structure, which adds to its total absorption over a wide frequency band.

Polymers loaded with NiZn ferrite with excess Fe [2, 7–9] will produce broadened absorption peaks, dependent on the NiZn ratio, for similar frequencies to the solid ferrite. The degree of absorption depends on the ferrite loading by volume and the grain size of the ferrite powder. The available permeability is significantly lower than the bulk ferrite, as is the dielectric constant, but for impedance matching purposes that can be an advantage at UHF and microwave frequencies. For some applications where high permeability is required at frequencies of about 1 MHz for maximum inductance, this can be a limiting factor, although permeabilities of about 200 are possible with optimum loading and suitable very high Zn/Ni ratio NiZn powder compositions.

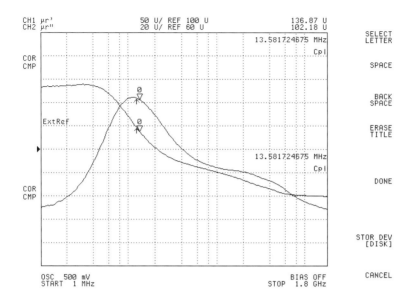

Figure 3.1   NiZn ferrite absorber.

Thin films of NiZn ferrites, produced by chemical deposition, reactive sputtering, laser ablation, and so forth can produce material with very different absorption characteristics [10]. This can be because of iron excess stoichiometry, leading to Fe^{+2} formation, or stress induced anisotropy. Such films have potentially more absorption per unit thickness than conventionally prepared ferrite. Co^{+2} can be used to extend these effects to even higher frequencies [11].

Noise filters and RF absorption are the subject of current research as one type of application is similar to a lowpass filter, passing signals up to a certain critical frequency band, and absorbing signal by magnetic loss above that frequency band. For ideal noise suppression, the magnetic loss should then fall as the reciprocal of the frequency. The possible methods of achieving this objective at higher frequencies using magnetic absorption are shown in Table 3.2 for materials of approximately the same low-frequency (initial) permeability. The data is from [7], plus standard solid ferrite data for commercially available material. A magnetic alloy powder (Fe-Si-Al metal)–filled rubber is included for comparison, also from [7].

Table 3.2
Frequency Characteristics of Bulk and Film NiZn Ferrite

Material	Initial Permeability, μ	Frequency Where Permeability falls (MHz)	Maximum Passband Frequency (MHz)	Range of Absorption Frequency (MHz)
Fe-Si-Al metal/ rubber	54	10	5	5 to 3,000
NiZnFerrite Film on Plastic	48	100	60	60 to > 10,000
Bulk NiZn Ferrite	80	10	5	10 to 12,000
NiZn film on Transmission Line	35	500	200	200 to > 10,000

Nickel zinc ferrites of the type shown in Table 3.2 can be fabricated by direct deposition of stressed NiZn spinel films with excess Fe, nominally as $(NiZn)_{0.5}(Fe^{+2})_{0.5}Fe_2^{+3}O_4$ films. These pass frequencies much higher than obtained by conventional means, up to 500 MHz. They are reported to be successful absorbers following the 1/frequency characteristic, but with much higher magnetic losses than magnetic metal/rubber sheet, such that power absorption is up to 20 times greater for the same film or sheet thickness. These films open up the possibility of micron scale absorbers for high-speed digital and microwave microelectronic packages, which currently use either much bulkier package scale absorbers or Faraday cage metal enclosures to prevent inter device radiation. Very thin absorbers are increasingly required because of package height restrictions even for larger-scale circuits.

3.3 Water as an Absorber

Perhaps surprisingly, water is used as an absorber in loads in the microwave heating industry. Water undergoes a classic Debye dielectric relaxation process at microwave frequencies and can be modeled as such to predict its behavior over frequency and temperature [12–14]. Water has a rapidly changing dielectric constant and loss with temperature, but can be used to determine power absorption when used as a load by flowing water calorimetry. At a low temperature, close to its freezing point, water still has a very high dielectric constant, but upon freezing to form ice, which has a very low dielectric constant (about 3) and low loss, there is an abrupt change in dielectric constant and loss behavior (Figures 3.2 and 3.3).

Water shows its classic Debye relaxation at 10 GHz, as the dielectric constant falls off rapidly, corresponding to the peak in absorption shown in Figure 3.3.

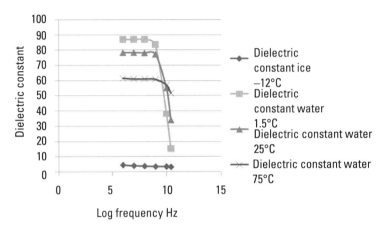

Figure 3.2   Dielectric constant (Y-axis) of ice and water versus log frequency, Hz. (*After:* [15].)

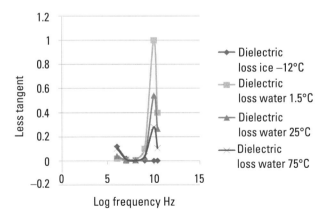

Figure 3.3   Dielectric loss of ice and water (Y-axis) versus log frequency, Hz. (*After:* [15].)

The peak in water's RF absorption can be seen at 10 GHz, at its highest just above freezing, whereas ice has very low dielectric loss above about 10 MHz.

These factors are very important in the food, textile, and paper industries where microwave heating and drying are used extensively, to say nothing of domestic microwave cooking, where frozen food's heating properties are often determined by the temperature and content of its enclosed water or ice.

In outdoor situations, the behavior of water as a surface and as an attenuator in the form of high humidity or rain in the atmosphere is also important for antennas and round trip loss calculations in radio systems.

Absorbed water in porous ceramics or composite materials is an obvious issue in microwave materials selection, and great caution is required when working with materials with open porosity. Composites are even more susceptible to water ingress because of the ceramic plastic interface, including printed circuit boards made of glass fiber and polymer.

The role of water vapor in air should also be kept in mind when dealing with closed metal cavities or waveguides. The Q of the air itself is reduced by the presence of water vapor, and also small differences in temperature between surfaces can lead to condensation on cooler surfaces if vapor saturation conditions exist. Flash condensation of very thin films of water will cause large drops in Q on high-Q dielectric materials. Where there is continuous temperature cycling in humid external conditions, closed metal high-Q structures should have either near hermetic sealing in dry air or nitrogen, or be fitted with drainage holes.

3.4 Barium Titanate Piezoelectrics

Barium titanate shows a relatively narrowband absorption peak at a relaxation frequency in the 1-GHz region (Figures 3.4 and 3.5). The dielectric constant falls precipitously in this region from the thousands to hundreds [16].

Absorbers of this type are only useful where a very dielectric constant or capacitance is required at the low-frequency, low-loss side of the relaxation peak, and a relatively low dielectric constant or

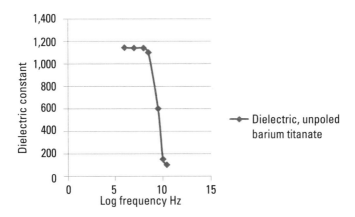

Figure 3.4 Dielectric constant (Y-axis) of unpoled barium titanate versus log frequency, Hz. (*After:* [15].)

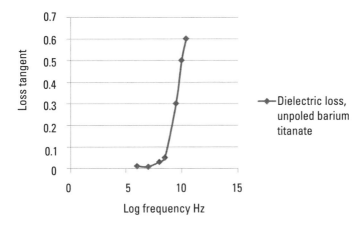

Figure 3.5 Dielectric loss of unpoled barium titanate versus log frequency, Hz. (*After:* [15].)

capacitance at the high-frequency side, with a corresponding high absorption (Figure 3.5). Note that these values are for unpoled piezoelectric barium titanate. In principle, all ferroelectric materials will show this type of behavior, including barium strontium titanate, known as BSTO or BaSrTiO$_3$ at relatively low levels of Sr (Chapter 11). When sufficient Sr is added to suppress the ferroelectric Curie temperature below operating temperatures, these materials behave as a paraelectric over relatively small temperature and compositional ranges, and have voltage-tunable dielectric constants and hence capacitance, albeit with very modest Qs.

In principle, voltage-tunable absorbers could be made in the same way, operating below the Curie point.

3.5 Silicon Carbide Absorbers

Silicon carbide is an intrinsic semiconductor and can be both n- and p-type by suitable doping. It occurs in both the beta form (cubic zinc-blende) or the alpha (hexagonal) form depending on the manufacturing conditions. The microwave properties are similar for both types. Silicon carbide is strong, has low expansion (~4 ppm/°C), high thermal conductivity (0.14 cal/cm-sec. °C) and can operate at high temperature (up to 1,200°C), making it good for very high–power applications. Its main disadvantage is difficulty in machining because of its hardness (2,900 Knoop). Commercial material may be made as silicon carbide bonded with alumina or silicon nitride. Its high dielectric constant (Table 3.1) can make matching difficult and microwave waveguide loads are typically stepped or tapered to achieve reasonable matching over wide frequencies [17]. Ceramic silicon carbide foam materials exist with a lower dielectric constant. The dielectric properties of powder SiC were investigated by Baeraky [18] at 615 MHz, 1.412 GHz, 2.214 GHz, 3.017 GHz, and 3.82 GHz, at temperatures from 25 to 1,800°C. The powder packing density is not mentioned, but the dielectric constant of the powder, packed into tubes to form a series of TM resonators (Chapter 9) at the above frequencies, was about 6 at room temperature at all frequencies. This rose to about 8 at 1,300°C, and 11 at 1,800°C. The dielectric loss, which varied continuously from about 0.3 at 615 MHz to 0.65 at 3.8 GHz at room temperature, and reached a minimum value of about 0.1 at 615 MHz and 0.4 at 3.8 GHz at 500°C, returned to its room temperature values at 1,400°C. Above 1,400°C, values continued to rise to a dielectric loss of about 7 at 1,800°C as conductivity rose.

3.6 Magnetic Metal Polymer Composite Materials

These types of materials have grown in importance, as microwave absorbers are also useful for suppressing high-speed data-created noise, as data rates rise. Most such absorbers were originally made of iron or iron alloy dust and were incorporated into polymers to form absorbers, which can be molded or cast into appropriate shapes using thermoplastic or more frequently thermosetting polymers. There is a continuum of materials from those with very low binder volume, around 10%, which are used from the kilohertz to low megahertz region, depending on the type of powder and its particle size, to RF absorbers with as much as 50% polymer. For spherical particle shapes, the Kornetski [19] equation can be applied using the initial permeability of the metal and the polymer (binder) content to derive the low frequency permeability. The losses at low frequency are determined by the Legg equation [20] and include eddy current losses, which become significant at high metal-filling factors. For iron powder this will produce permeabilities in the range > 250 to << 10 depending on the fill factor. This is significant because the lower metal powder materials with very fine particle size have useful permeability and low losses up to about 100 MHz then become useful RF absorbers in the UHF and microwave region, with the carbonyl iron process the most ubiquitous. From a matching point of view, the relatively low permeability is useful but relatively high dielectric constants can be an issue. The change in dielectric constant, permeability, and magnetic and dielectric loss with volume filling fraction is shown in Figure 3.6.

Both dielectric constant and permeability rise with iron content, but the effect on dielectric constant is much greater at this frequency (0.1 MHz). There is a correspondingly greater rise in dielectric loss (Figure 3.7) compared with magnetic loss.

There are very large changes in surface and volume resistivity as iron is added (Figure 3.8), which should be kept in mind if the material is used at power, where breakdown may occur at relatively low levels. Silicon carbide may be a better choice for applications of more than a few watts of RF power.

The change in dielectric and magnetic properties of iron-filled 2,5 dichlorostyrene were first reported by von Hippel at much higher

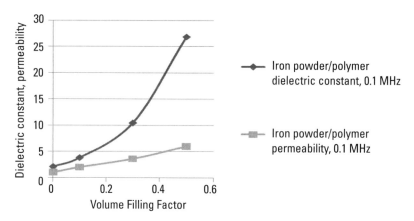

Figure 3.6   Permeability and dielectric constant (Y-axis) versus volume filling fraction for iron/polymer powder. (*After:* [9].)

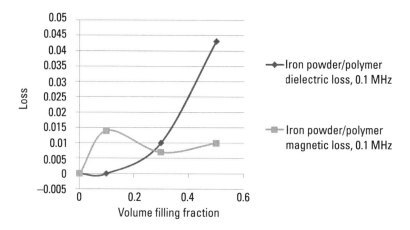

Figure 3.7   Dielectric and magnetic loss (Y-axis) versus volume-filling fraction for iron powder/polymer. (*After:* [9].)

frequencies, and the effect on dielectric constant for approximately 0.3 volume filling factor is shown in Figure 3.9.

The effect on magnetic loss at high frequency (Figure 3.10) can be seen to be much greater than at low frequency, reversing the low-frequency effect in Figure 3.7.

Figures 3.9 and 3.10 show the fall in dielectric constant and permeability of iron powder and polymer mixtures in the 10-MHz to

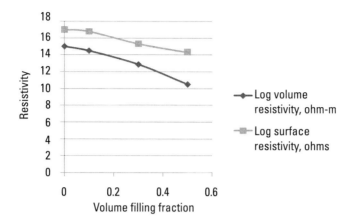

Figure 3.8 Log volume and log surface resistivity (Y-axis) versus volume-filling factor for iron powder/polymer. (*After:* [9].)

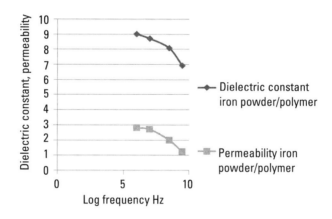

Figure 3.9 Dielectric constant and permeability (Y-axis) of iron powder/polymer versus log frequency, Hz. (*After:* [15].)

3-GHz region, accompanied by a modest rise in dielectric loss and a very large increase in magnetic loss (Figure 3.10), making this series of materials very useful as bulk absorbers in the microwave region at low RF power levels.

A large range of options for metal-filled polymers opens up when other metal alloys are used with higher permeability or higher resistivity [21], or different particle shapes are used [22]. This can raise the

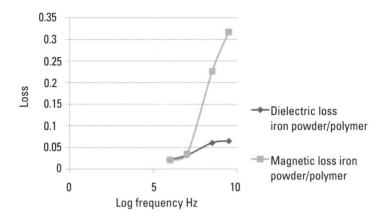

Figure 3.10   Dielectric and magnetic loss (Y-axis) of iron powder/polymer versus frequency, Hz. (*After:* [15].)

permeability of a metal polymer composite for a given volume-filling factor, or if a flake shape is used, an oriented sheet of composite can be created with anisotropic magnetic and dielectric properties. Eddy currents in flake-shaped particles can be used to suppress low-frequency noise while the composite retains higher-frequency absorption capability from the magnetic relaxation effect. Eddy currents are limited by the skin depth effect at higher frequencies. In flake-filled polymer sheets, layers of flakes may be insulated from each other by the polymer, creating very different magnetic and dielectric properties in the Z-direction compared with properties in the X-Y plane. In general, however, it is possible to create materials with low losses below 1 MHz, with a wide range of permeability and hence inductance, with excellent RF absorption capability from 1 MHz upwards, comparable with solid ferrite. Ferrite fillers have much higher resistivities than metal, and eddy currents do not occur. Reduced ferrite fillers containing large amounts of Fe^{+2} do have high dielectric losses because of electron hopping. For a complete discussion on noise suppression materials and applications, see [23].

Thin-film noise suppression may ultimately take the form of thin films of ferromagnetic alloys or compounds [24] for microelectronic packaging.

3.7 Hexagonal Ferrite Absorbers

Hexagonal ferrites, which retain permeability well above the Snoek limit, are obvious candidates for high-frequency absorbers above their relaxation frequency, either as solid ferrite or as ferrite powder in polymers. Z-types can be used in the UHF range [25] and above, and Y-types can be used in the microwave region [26]. Selective absorbers can be made with M-type hexagonal ferrites [27], which have relatively narrowband absorption. In polymer elastomers, it is possible to combine several types of absorber into a single composite. (See also Chapter 4, Section 4.8 on composites.)

3.8 Summary

In general, for low-power applications at microwave frequency, iron powder-filled polymers are used as absorbers. Ferrite-filled polymers and bulk ferrite are used where high permeability and lower loss is required at higher frequencies in the megahertz range. Bulk ferrite or hexagonal are used where higher dielectric constants and permeability are required up to hundreds of megahertz. Silicon carbide is used for high-power and high-temperature absorbers. Water and ferroelectrics are used for specialized applications. Very thin films of deposited ferrites may become useful for microelectronic microwave use.

References

[1] Goodenough, J. B., "Summary of Losses in Magnetic Materials," *IEEE Trans. Mag.*, Vol. 38, No. 5, 2002, p. 3398.

[2] Dionne, G. F., "Magnetic Relaxation and Anisotropy Effects on High Frequency Permeability," *IEEE Trans. Mag.*, Vol. 39, No. 5, 2003, p. 3121.

[3] Globus, A., and M. Guyot, "Control of the Susceptibiliy Spectrum in Polycrystalline Ferrite Materials and Frequency Threshold of the Losses," *IEEE Trans. Mag.*, Vol. MAG-6, No. 3, 1970, p. 614.

[4] Gieraltowski, J., and A. Globus, "Domain Wall Size and Magnetic Losses in Frequency Spectra of Ferrites and Garnets," *IEEE Trans. Mag.*, Vol. MAG-13, No. 5, 1977, p. 1357.

[5]   Pyun, S. I., and J. T. Baek, "Microstructural Dependence of Permeability and Permeability Spectra in Ni-Zn Ferrites," *Amer. Ceram. Soc. Bull.*, Vol. 64, No. 4, 1985, p. 602.

[6]   Valenzuela, R., and J. T. S. Irvine, "Domain Wall Relaxation Frequency and Magnetocrystalline Anisotropy Constant in Ni-Zn Ferrites," *J. Mag. Mat.*, Vol. 160, 1996, p. 386.

[7]   Matsushita, W., and M. Abe, "Conducted Noise Suppression in GHz Range by Composite Sheets and Ferrite-Plated Films," *ICF-9*, 2004, p. 739.

[8]   Kim, S. S., et al., "Complex Permeability and Permittivity and Microwave Absorption of Ferrite-Rubber Composite at X-Band Frequencies," *IEEE Trans. Mag.*, Vol. 27, No. 6, 1991, p. 5461.

[9]   Shin, J. Y., and J. H. Oh, "The Microwave Absorbing Phenomena of Ferrite Microwave Absorbers," *IEEE Trans. Mag.*, Vol. 29, No. 6, 1993, p. 3437.

[10] Kondo, K., et al., "GHz Conducted Noise Suppression by NiZn Ferrite Films Plated onto Polyimide Sheets," *ICF-9*, 2004, p. 659.

[11] Matsushita, N., and M. Abe, "Ni-Zn-Co Ferrite Films Prepared at 90 Degrees C Having Permeability of 30 at 3GHz," *IEEE Trans. Mag.*, Vol. 38, No. 5, 2002, p. 3111.

[12] Stogryn, A., "Equation for Calculation of the Dielectric Constant of Saline Water," *IEEE Trans. MTT*, 1971, p. 733.

[13] Klein, L., and C. Swift, "An Improved Model for the Dielectric Constant of Sea Water at Microwave Frequencies," *IEEE Trans. Ant. Prop.*, Vol. AP-25, No. 1, 1977.

[14] Grant, E., T. Buchanan, and T. Cook, "Dielectric Behaviour of Water at Microwave Frequencies," *J. Chem. Phys.*, Vol. 26, 1957, p. 156.

[15] Von Hippel, A., *Dielectric Materials and Applications*, New York: John Wiley/ MIT, 1954.

[16] McNeal, M. P., S. J. Jang, and R. E. Newnham, "The Effect of Grain and Particle Size on the Microwave Properties of Barium Titanate," *J. Appl. Phys.*, Vol. 83, No. 6, 1998, p. 3288.

[17] Matsumoto, H., et al., "Experience on the High-Power SiC Microwave Dummy-Load Using SiC Absorber," *Proc. Conf. Particle Accelerator*, 1999, p. 842.

[18] Baeraky, T. A., "Microwave Measurements of the Dielectric Properties of Silicon Carbide at High Temperature," *Egypt. J. Sol.*, Vol. 25, No. 3, 2002, p. 263–273.

[19] Kornetski, M., and H. Weis, "Theoretical Calculation of Permeability of Iron Powder Pressings," *Siemens Werken*, Vol. XV, p. 95.

[20] Legg, V. E., and F. J. Giuen, "Relationship Between Metallic Packing Factor and Permeability in Iron Dust Cores," *J. Bell Systems Tel.*, Vol. 19, 1940, p. 385.

[21] Gokturk, H. S., T. J. Fiske, and D. M. Kaylon, "Electric and Magnetic Properties of a Thermoplastic Elastomer Incorporated with Ferromagnetic Powders," *IEEE Trans. Mag.*, Vol. 29, No. 6, 1993, p. 4170.

[22] Golt, M., "Magnetic and Dielectric Properties of Magneto-Dielectric Materials Consisting of Oriented, Iron Flake Filler Within a Thermoplastic Host," Master's Thesis, University of Delaware, 2008.

[23] Yamaguchi, M., and K. H. Kim, "Noise Suppression Sheet/Film for Digital Devices and Equipments," *IEEE MTT-s*, IMS, 2005.

[24] Ohnuma, S., et al., "Metal-Oxide Type Nano-Granular Soft Magnetic Films and Their Noise Suppression Effects at GHz Frequencies," *ICF-9*, 2004, p. 775.

[25] Martin, L. J., et al., "On Components and Packaging Technologies," *IEEE Trans. Comp. Pkg. Tech.*, Vol. 32, No. 4, 2009, p. 849.

[26] Kwon, H. J., J. Y. Shin, and J. H. Oh, "The Microwave Absorbing and Resonance Phenomena of Y-Type Hexagonal Ferrite Microwave Absorbers," *J. Appl. Phys.*, Vol. 75, No. 10, 1994, p. 6109.

[27] Nedkov, I., A. Petkov, and V. Karpov, "Microwave Absorption in Sc- and CoTi-Substituted Ba Hexaferrite Powders," *IEEE Trans. Mag.*, Vol. 26, No. 5, 1990, p. 1483.

4

Plastics and Plastic Ceramic Composite Materials

4.1 Introduction

Microwave dielectrics can be divided into categories essentially by their dielectric constant. If we ascribe the dielectric constant at the molecular level to polarizability of the chemical bond between the key elements in the structure, we find that of the two major categories of dielectrics, plastics have low polarizability and oxide ceramics have a higher and wider range of values. There is a range of values where an alternative category may be the best choice, namely composites of ceramic and plastics.

When considering dielectric loss, it is possible to indentify groups of materials which are likeliest to give very low dielectric loss at microwave frequencies, based on aspects of their chemical composition. In general, this chapter will begin with the lowest dielectric constants and finish with the highest, with comments on the commercial availability and actual performance of typical materials.

4.2 Plastics and Hydrocarbon Polymers

Plastic polymers are based on long carbon chains. Because of the unique nature of the carbon atom, typically two of its four available covalent bonds connect each member of the carbon chain, and two are available to form bonds with other atoms or groups of atoms through a single bond, which may be of sufficient complexity to form a short side chain. In practice, the main chain may be branched, or there may be more than one type of chain (copolymers). The chain itself has a three dimensional or tertiary structure in space, which if ordered sufficiently is described as crystalline, or if disordered as amorphous. In some types of chains, some carbon atoms may be replaced by others, such as oxygen.

A further categorization of plastics is their ability to be formed into shapes. There are three types. The first is thermoplastics, made as powder or slugs, which can be melted then injection molded into complex shapes, but which are difficult to fill with powder. The second is thermosetting plastics, which are typically formed by a chemical reaction when poured into shaped molds. The third is similar to thermoplastics but cannot be melted, and thus have to be formed by die pressing, extruding or related methods, followed by sintering to a dense part.

The properties listed for various plastics are based on typical values for classes of polymers available in the technical or commercial literature. In practice, some commercial plastics may not be exactly the same because of differences in polymer chain length, and the use of co-polymers and other additives used in their manufacture. Nevertheless, the values chosen are intended to be the most representative of each class of polymer when used in a microwave application.

4.2.1 Hydrocarbon-Based Polymers

In Chapter 1, the concept of valency was discussed. In oxides like ferrites, electrons given up by a metallic cation are shared with the oxygen atoms. Although spin interactions like superexchange occur, for all practical purposes they remain so, creating a strongly polarized metal to oxygen bond, mainly Fe^{+3} to O^{-2} in ferrites, and hence in the presence of an electric field a relatively high dielectric constant, in the range 12 to 18 for spinels and garnets. This type of bond is called covalent and also occurs in polymers, the difference lying in the degree of polarization, or ionic content. Carbon is tetravalent, that is it can share four electrons, but if these are all with monovalent hydrogen, the degree of polarization of the C to H bond is very small and hence the dielectric constant of a solid containing these bonds is very low, about two. In the simplest of hydrocarbons polymers, the solid consists of a chain of carbon to carbon bonds, with each carbon atom also bonded to two hydrogen atoms. The conventional way to represent the chain is $-(CH_2)-(CH_2)-(CH_2)-(CH_2)-$ and so on, abbreviated to (CH_2) n where n can be thousands or millions of repeating units. For copolymers where more than one type of chain is present, n, m, p, and so forth represent the number of repeat units per chain.

If we look at the most simple type of hydrocarbon polymer, where there is either no side chain (polyolefins like polythene) or a short hydrocarbon chain with single bonds (polypropylene and 1,5 dimethyl pentene polymers like TPX), these have intrinsically very low polarization. Low polarization means that relaxation mechanisms are unlikely as the electromagnetic frequency increases, and hence dielectric loss is likely to be low at all frequencies in the spectrum. As a result, we see dielectric constants of around 2 and potential dielectric losses in the region of 0.0001 at microwave frequencies. As the side chain lengthens, the plastic has an increasingly higher melting point and rigidity, making TPX, for example, extremely useful as a microwave plastic. This group of hydrocarbon polymers is thermoplastic.

4.2.2 Hydrocarbons with Aromatic Side Chains

The simplest form of this is polystyrene, which has a benzene ring type of side chain. The convention for representing an aromatic ring like benzene is simply R, so a polymer like this can be represented as $(CHR)n$. The benzene ring is actually a six-member carbon ring, with each carbon having two hydrogen bonds, C_6H_{12}. Although the electrons are delocalized in the ring, the bond is essentially covalent and low dielectric losses are found. Polystyrene can be made in foam form by suitable foaming agents during the forming process, giving a wide range of very low dielectric constants from slightly more than 1, to 2.55. Higher-temperature polystyrene types are possible by substitu-

Table 4.1
Hydrocarbon-Based Polymers: Dielectric and Thermal Properties

Polymer	Trade Name	Chemistry	Dielectric Constant/ GHz	Dielectric Loss/ GHz	Melting Point (°C)	Vicat Deflection Point (°C)	Comments
High-density polyethylene	Polythene	(—CH2—)n	2.37 at 11.3	0.0001 at 11.3	Variable ~100	80 + C	
Polypropylene		–(CH2CHCH3)n	2.26 at 9.4	0.0001 at 9.4	160	130	
Poly4-methyl pentene	TPX	(CH2CHCH2CH2CH2CH2CHCH3CH3)n	2.1 at 12	0.0008 at 12	236	185	Density 0.83 g/cc
Polystyrene		(CH2CHC6H6)n	2.55 at 9	0.0003 at 9	100	90–100	
Cross-linked polystyrene	Rexolite	Polystyrene/ divinylbenzene crosslink	2.53 15 MHz to 50 GHz	0.00066 at 10.0	100+	90–100	+100°C max temperature

tion in the aromatic ring or by more complex side chains. A variant of hydrocarbon-based polymers are the cross-linked polyolefins (Rexolite, for example), which are aromatic polyolefins that are thermosetting by initiating cross-linking after molding, and as a result can also readily be filled. Hence, these are one of the few low-loss plastics groups that can be filled, because other thermosetting plastics have significantly more loss dielectrically. Of this group only TPX and polypropylene will operate continuously without some degradation above 100°C, a relatively low temperature for many system requirements, and none would meet the needs of solder reflow temperatures used on surface-mounted components on printed circuit boards except for very short duration exposure. The characteristics of each hydrocarbon type are given in Table 4.1. The data comes from [1–3] plus manufacturers' literature. Methods of measurement are included in [1, 2, 4]. Note that both melting point and Vicat deflection point are given wherever possible. Where there is a glass transition point, this is given instead of the melting point as some polymers do not truly melt, are amorphous, or may decompose without melting.

4.3 Fluorocarbon-Based Polymers

Fluorine forms covalent bonds with carbon and as a result a gradation of substitution of fluorine for hydrogen is possible. The best known example is Teflon or polytetrafluoroethylene (PTFE), which has a low dielectric constant and very low loss at microwave frequencies. The main disadvantage is its anomaly in thermal expansion near room temperature caused by a phase transition, and a resulting tendency to creep when compressed or confined mechanically. Like many unfilled polymers, it is difficult to achieve good adhesion when gluing, and surface treatment may be necessary. It also cannot be injection molded or extruded as it is not a thermoplastic. Thermoplastic fluorocarbons can be made, for example as copolymers of tetrafluoroethylene with perfluoropropylene producing material like FEP, or higher temperature–resistant PFA (perfluoroalkoxy polymer). In general, substitution of other atoms for fluorine or hydrogen, particularly oxygen, will produce higher dielectric losses in these polymers, but PFA is an exception at least up to 10 MHz, although it may become higher loss at microwave

Table 4.2
Fluorocarbon Polymers: Dielectric and Thermal Properties

Polymer	Trade Name	Chemistry	Dielectric Constant	Dielectric Loss	Melting Point	Maximum Use Temperature	Comments
PTFE	Teflon	(C2F2)n	2.04–2.08/11.5	0.0008/11.5 GHz	327 phase change	260	Does not melt
FEP		(C2F2)n + (C2F.CF3)m	2.02/10.4	0.0008/10.4 GHz	260	204	
PFA		(C2F2)n + (CF2.CF.OCF3)m	2.1/10 MHz	0.0003/10 MHz	305	260	

Table 4.3
High Temperature Unfilled Plastics: Dielectric and Thermal Properties

Polymer	Trade Name	Chemistry	Dielectric Constant	Dielectric Loss	Use Temperature	Melting Temperature	Comments
Polyimide	Kapton	—	2.6–3.2/ 1 to 110 GHz	0.006/ 1 to 110 GHz	Vicat 220°C	> 400°C	Tape form
Polysulphone	Udel unfilled	—	2.98/10.8 GHz	0.01/10.8 GHz	Heat deflection 174°C	Glass transition 185°C	—
Polyethyl ketone	PEEK unfilled	—	~3	0.004/10 GHz	Glass transition 143°C	343°C	—

frequencies. These materials will withstand higher temperatures than polyolefins and have very low dielectric loss, but are still relatively soft. The properties of typical fluorocarbons are shown in Table 4.2 [1–3].

4.4   Structural Thermoplastics

Other thermoplastics are used at microwave frequencies for structural elements, where mechanical strength and resistance to creep at high temperatures are potential issues and dielectric loss is less important. These include polyimides, polysulphones, polyetheretherketone (PEEK), and polycarbonates. These have base dielectric constants around 3 when unfilled and relatively low loss tangents, such that they can be used for structural application without increasing losses too severely. They are typically filled with reinforcing glass in fiber or mat form to improve their basic mechanical and thermal characteristics, but Table 4.3 lists their unfilled properties to allow a proper comparison of the polymers only. Many of these polymers are hydrophilic, in contrast to the polyolefins and fluorocarbons, which are hydrophobic. This often means the former are easy to fill but are subject to the ingress of moisture, whereas the latter are difficult to fill but water repellant.

One of the best known of this thermoplastic group is the polyimide polymers. The most used are aromatic heterocyclic imides. These gain their very good mechanical and thermal strength from chain-to-chain donor acceptor bonding between the nitrogen atoms of the heterocyclic ring (N and C in the same ring) and the double-bonded (to carbon) oxygen atoms in the chain itself, giving a three-dimensional rigidity to the structure. These bonds, however, are also likely to be the main source of the higher dielectric constant and loss of these materials at microwave frequencies relative to the hydrocarbon and fluorocarbon-based polymers. Nevertheless, the unfilled material is extremely useful in tape form (Kapton) and in the glass filled form as a bulk plastic (Ultem). In principle polyimides can either be thermosetting or thermoplastic, although the latter requires special preparation. The main disadvantage of polyimides is their susceptibility to humidity and moisture absorption because of the condensation reaction required to form them. These can cause very large increases in dielectric constant and loss in extreme environments and may be unusable if not suitably protected.

Other high-temperature plastics include polyethylsulphones (Udel) and polyetheretherketone (PEEK). Polysulphones have four benzene rings connected to each other by a dimethyl carbon link, an alternating oxygen link, and a sulphone link per monomer unit. Polyethyl ketones have three benzene rings with alternating CO and oxygen links. These linkages are likely responsible for the higher dielectric constants and losses of these materials as well as their improved thermal properties. PEEK in particular is capable of surviving lead-free soldering temperatures, although normally mechanically stiffened by glass filler. The higher-temperature polymer properties are summarized in Table 4.3. Sources include [1–3] and manufacturers' literature. Polyimide has been measured from 1 to 110 GHz [5].

For completion, the microwave properties of cheaper but less electrically and thermally effective polymers are included (Table 4.4) as these are sometimes used for structural elements in antennas, radomes, and RF enclosures in benign environments [1].

4.5 Epoxies

This group of thermosetting plastics covers a wide range of laminates, casting compounds, and glues. In general, these do not have low dielectric loss but are useful for some microwave applications. As the base for most applications, there is a class of epoxies usually made from bisphenol A and epichlorohydrin, followed by crosslinking at the curing stage with an amine or anhydride catalyst. This type has been reported to have dielectric constants in the range 3.5 to 3.7 and losses of 0.04 to 0.07 over 10 to 25 GHz. Specific measurements have been on epoxies at 915 MHz and 2.45 GHz [6] and also 22 GHz [7]. However, by changing the many types of possible combinations of bisphenol and related compounds, or the use of dicyclopentadiene (DCPD) in development epoxies, the dielectric constant in future materials may be as low as 2.5 to 2.7 and the losses may be 0.001 to 0.005 for the 10- to 25-GHz range. Most epoxies degrade above 180°C and soften about 50°C lower than that temperature, but advanced grades can soften above 200°C and withstand short term exposure above 260°C.

Table 4.4
Lower-Cost Plastics: Dielectric and Thermal Properties

Polymer	Typical Name	Chemistry	Dielectric Constant	Dielectric Loss	Heat Resistance	Melting Point	Comments
PMMA	Perspex, Lucite	(C5O2H8)n	2.61/11 GHz	0.008/11 GHz	Vicat 110°C	160°C	—
PVC	PVC	(CH2CHCl)n	2.70/11 GHz	0.008/11 GHz	Glass transition 87°C	~200°C	—
Polyamide	Nylon 66	Polyamide with C4,C6 alkanes	3.03/ 10.8 GHz	0.01/10.8 GHz	~100°C	265°C	Moisture sensitive
ABS copolymer	Cycolac	(C8H8)n+ (C4H6)m+ (C3H3N)p	2.79/11 GHz	0.008/10.8 GHz	Vicat 99°C	Glass transition 105°C	—
Acetal homopolymer	Delrin		2.73/11 GHz	0.007/11 GHz	Deflection 97°C	178°C	—
Polycarbonate	Lexan	Bisphenol A	2.77/11 GHz	0.005/11 GHz	Vicat 145°C	267°C	—

Because of their importance as laminates for high-frequency printed circuit boards and structural elements in microwave applications, and as resin for powder filled absorbers, the filled properties of these resins will be discussed in detail in Section 4.8.

4.6 Silicones

This group of thermosetting materials is used for casting, gluing, and as a resin for flexible absorbers. Basic silicones such as polymethyldiethylsiloxane are reported to have a dielectric constant of just below 4 and a loss tangent of about 0.025, at 1 GHz. For most microwave applications they are used with fillers, but some are used as glue or tuning materials without fillers because of their high-temperature characteristics, as most silicones will operate above 200°C. Some glue types are available (GE or Dow) with alumina-based fillers to improve thermal conductivity while retaining reasonably low dielectric loss. These are used on microwave ceramic to ceramic bonds where there is some mismatch in thermal expansion. Again, joint preparation is critical for adhesion as the filled silicone bond is not as strong or moisture resistant as epoxy glue bonds.

As glues, silicones are extremely important for joining materials with different expansion coefficients together. While epoxies can be used for joining closely matching materials, within 1–2 ppm/C, silicones, because of their flexibility, can join materials with mismatches of 10 ppm/C or more. Because silicones in general form weaker bonds than epoxies, much more attention needs to be paid to surface cleaning, use of chemical treatment to enhance bonding, bond thickness, and thermal treatment to optimize the bond. Thermal expansion problems may not reveal themselves until many temperature cycles occur, usually created by environmental or RF heating effects, so accelerated life testing may be required to prove long-term reliability.

Silicones are commonly used as the polymer base for absorbers (Chapter 3) containing ferrite or magnetic metal powder, so their dielectric properties are secondary to their flexibility and temperature stability. They are typically used in sheet form. Some indication of their performance up to 1 GHz is available in [8].

Table 4.5
Dielectric Properties of Glasses at 3 and 10 GHz at Room Temperature (RT) and 100°C

Glass Type	Oxide Composition	Dielectric Constant 3 GHz	Dielectric Constant 10 GHz	Dielectric Loss 3 GHz	Dielectric Loss 10 GHz	Notes
Fused silica room, RT	Si	3.78	3.78	0.00006	0.0001	
100°C		3.78	3.78		< 0.0001	
Borosilicate RT	Si, B, low alkali	4.05	4.05	0.0011	0.0016	
100°C						No change
E-glass RT	Si, Al, Mg, Ca, low alkali		6.11		0.006	
100°C						Will rise substantially
High alkali	Si, K, Na		5.4		0.012	
100°C			5.5		0.018	

Table 4.6
Density and Dielectric and Thermal Properties of Common Fillers

Filler Dielectric	Dielectric Constant 10 GHz	Dielectric Loss 10 GHz (max)	Temperature Coefficient Dielectric Constant ppm/°C	Density g/cc	Thermal Conductivity Cal/cm²/cm/sec/°C
Alumina	9.9	0.0001	+115	3.9	0.045
Barium tetratitanate	37	0.0005	–25	4.4	0.01
TTB titanates	78–90	0.001	0	5.7 to 6+	0.01
Titania	100	< 0.001	-575	4	0.01
Calcium titanate	170	0.002	–1,250	3.9	0.01
Strontium titanate	270	0.002	–2,200	5.1	0.01

4.7 Polyurethanes

These are typically used as high-loss foams when filled with carbonyl iron, carbon, ferrite, and so forth to act as absorbers, and the basic dielectric properties are not of any direct use. Their main advantage is flexibility as they are less temperature stable than silicones.

4.8 Filled Polymers

There are an enormous number of filled polymer composites. However, there are certain features that cannot be changed by adding any inorganic filler. These include the melting point or glass transition of the polymer, although the softening point may be raised. The dielectric properties are changed by the volume fraction of the filler, and can be calculated by the "log mixing rule," for example. However, as can be expected, no matter how low loss the filler, the overall loss of the composite is determined mostly by the polymer, suitably diluted by the filler, except at very high load factors.

4.8.1 Types of Fillers

For reasonably low-loss dielectrics, the purpose of the filler is either to raise the dielectric constant or improve its thermal or mechanical properties. Fiber-based fillers, usually types of glasses, are mainly aimed at controlling dimensional stability. This may be three-dimensional in the case of simple short fibers, or two-dimensional in the case of a woven mat. Each class of polymer will be considered in terms of ease of filling and type of fillers used.

Glass used for composites comes in several grades and varies in dielectric properties depending on its composition. Fused silica is a form of quartz and is very low loss, whereas borosilicate or soda glasses have higher losses. Hence the loss factor of a glass-filled low-loss polymer can be expected to be influenced by the grade of glass. The glasses were studied at microwave frequencies at MIT [2] many years ago but the data has stood the test of time. Basically the basis can be considered pure silica in the form of fused quartz, which has stable dielectric behavior over frequency and temperature. Borosilicate glass with very small amounts of alkalis or alkali earths has slightly higher dielectric

Table 4.7
Filled Polyolefin Hydrocarbon Dielectrics: Dielectric, Thermal, and Water Absorption Properties

Polymer	Filler	Dielectric Constant	Dielectric Loss	Expansion Coefficient ppm/°C	Maximum Use Temperature °C	Thermal Conductivity cal.cm/cm²/ sec/°C	Water Absorption%
Rexolite 1422	Unfilled	2.53	0.0005	70	90–100	0.00035	<0.05
Rexolite 2200	Glass fiber	2.62	0.0014	57	100	0.0005	0.1 max
Eccostock 0005	unfilled	2.53	0.0005	63	90–100 200 short term	0.0003	0.03
Eccostock HiK500F	titania	3 to 30 range midrange	<0.002	36 midrange	250 short term	—	"low"

Source: [9].

Table 4.8
Filled High-Temperature Polymers

Polymer Base	Filler	Dielectric Constant (GHz)	Dielectric Loss (GHz)	Thermal Expansion ppm/°C	Maximum Use Temperature (°C)	Water Absorption % 24 hr	Tg or Decomp. Temperature (°C)
PTFE Nelco N.9000	Woven glass range	2.08 to 2.33 at 10 GHz	0.0006 to 0.0011 at 10 GHz	X 25 Y 35 Z 260	>> 260	0.02	Decomp. 400+
PTFE Rogers RO3003	Glass	3.0 at 10 GHz	0.0013 at 10 GHz	X,Y 17 Z 24	>> 260	< 0.1	Decomp. 400+
PTFE Rogers RO3006	Glass + ceramic	6.15 at 10 GHz	0.002 at 10 GHz	X,Y 17 Z 24	>> 260	< 0.1	Decomp. 400+
PTFE Rogers RO3010	Glass + ceramic	10.2 at 10 GHz	0.0023 at 10 GHz	X,Y 17 Z 24	>> 260	< 0.1	Decomp. 400+
Epoxy FR4 Nelco N4000-2	Glass	4.1 at 1 GHz	0.023 at 1 MHz (non-RF)	X,Y 12–16 Z 450	~200 short term	0.1	Tg 140
Epoxy Nelco N.9350	Glass	3.7 at 10 GHz	0.004 at 10 GHz	X,Y 10–14 Z 48 to Tg	260 short term	0.15	Tg ~200
Epoxy Nelco 4350-13RF	Glass	3.5 at 10 GHz	0.0065 at 10 GHz	X,Y 10–14 Z 350 to 260C	260 short term	0.1	Tg 210
Polyimide Nelco N7000-1	Glass	3.8 at 10 GHz	0.0095 at 10 GHz	X,Y 12-15 Z 180 to260°C	350+	0.35	Tg 260
PEEK	30% glass	3.7 at 1 MHz	0.004 at 1 MHz	22	315	0.11	Tg 260
Poly sulphone	30%	4.2 at 1 MHz	0.007 at 1 MHz	27	216	0.35	Tg ~140

constant and loss, the latter rising slightly with temperature. E-glass, used in many commercial applications, is high in alkali earths but low in alkali. The alkali elements Na and K substantially raise the dielectric constant and loss, and also raise these effects with higher frequencies and temperatures. The alkali earth elements have less effect on dielectric constant but do raise loss substantially. Their properties are summarized in Table 4.5, including their performance at 100°C.

Ceramic fillers are added primarily to increase dielectric constant but also can improve temperature stability of the dielectric constant and thermal conductivity. Examples of the types of powder that can be used are given in Table 4.6. Their bulk properties as solids are given to show their effect on polymers, of course in proportion to their fill factor. Their expansion coefficients are all in the region of 10 ppm/°C, so they have a similar effect on polymer composite expansion, for the same volume fraction. These dielectrics, including tetragonal tungsten bronzes (TTB) are discussed in more detail in Chapters 5 and 6. Titania (TiO_2) and other high dielectric constant titanates are used in some dielectrics as fillers, but have very high temperature coefficients of dielectric constant (Chapter 6).

4.8.2 Filled Polyolefins

These are difficult to fill in the case of polyethylene and the corresponding aliphatic polymers. Most commercially available filled materials are based on crosslinked aromatic (polystyrene)-based polymers. Two well-known examples are glass-filled laminate (Rexolite) and titania-filled dielectric (Emerson and Cuming HiK) aromatic polymers. At high loading factors, water absorption becomes an issue and loss will increase. At a high temperature, the filler will give additional mechanical rigidity, but some degradation of the polymer can be expected and caution is advised. Typical examples are given in Table 4.7.

4.8.3 Filled Fluorocarbons

These are relatively uncommon for microwave applications in bulk with the exception of Fluorosint, a Teflon/natural quartz composite that has found occasional use where a slightly higher dielectric constant, low-loss material is required. Its relatively coarse microstructure limits its use. Bulk glass-filled Teflon does exist but is not used much. Glass fiber filled fluorocarbon laminates, clad with copper, are, however used extensively, either with or without ceramic powder to raise the dielectric constant. Laminates use microwave-grade glass and ceramic powder of the type listed in Table 4.8 to achieve dielectric constants of approximately 3, 6, and 10 with fluorocarbon polymers. These include special antenna grades for structural and low intermodulation requirements.

4.8.4 Filled High-Temperature Polymers

Glass is used extensively in high-temperature plastics to improve heat distortion. Because these plastics are relatively high loss, no special low-loss grades of glass are used. Examples of this type are glass filled polyimide (Ultem), polysulphone, and PEEK. These are used in relatively high–Q cavities, microwave enclosures, and structures by minimizing the amount of material in high E-field positions. Radomes used in rugged environments are also subject to extreme environmental stresses. These plastics are available in injected molded form, but due attention is required to anisotropy in dielectric and mechanical properties due to alignment of the glass fiber during forming. Table 4.8 lists these properties to show their relative performance to filled epoxy and fluorocarbon polymers. Again, the water absorption for filled polyimides, polysulphones (PES), and PEEK are significantly higher compared with filled polyolefin and fluorocarbon polymers. This, together with their intrinsically higher dielectric losses, makes them potentially difficult to use in humid or wet conditions.

4.8.5 Filled Epoxies for Laminates

The standard FR4 board used at lower frequencies is often too high loss for microwave applications [10]; special grades of epoxy and glass are used, for example by Nelco [11], although they are not as good as fluorocarbon-based laminate. Epoxy-based glues in glass fiber reinforced form are used as adhesive preforms to maintain low expansion and dimensional stability during curing.

Some comparisons of a range of types of glass-filled polymers at high temperature are included in [12], including dielectric properties of glass- and quartz-filled epoxies, silicones, and polyimides, measured at 9.35 GHz and up to 260°C.

4.9 Summary

A basic understanding of the chemistry of polymers gives an insight into their likely dielectric and thermal properties. Thermal properties of the polymer will still dominate even at high fill fractions of ceramic or glass materials, with the exception of expansion coefficients, but dielectric, magnetic, and absorbing properties can be modified substantially.

References

[1]   Riddle, B., J. Baker-Jarvis, and J. Krupka, "Complex Permittivity Measurements of Common Plastics over Variable Temperatures," *IEEE Trans. MTT,* Vol. 51, No. 3, 2003, p. 727.

[2]   Von Hippel, A., *Dielectric Materials and Applications,* Cambridge, MA: MIT Press, 1961.

[3]   www.microwaves 101.com/ENCYCLOPEDIA/substrates_soft.efm.

[4]   Olyphant, M., and J. H. Ball, "Strip-Line Methods for Dielectric Measurements at Microwave Frequencies," *IEEE Trans. Elec. Insul.,* Vol. E1-5, No. 1, 1970, p. 26.

[5]   Ponchak, G. E., and A. N. Downey, "Characterization of Thin Film Microstrip Lines on Polyimide," *IEEE Trans. Comp., Pkg. Tech.-Part B,* Vol. 21, No. 2, 1998, p. 171.

[6]   Zong, L., L. C. Kempel, and M. Hawley, "Dielectric Properties of Polymer Materials at High Microwave Frequency," *IEEE Ant. and Prop. Soc. Int. Sym.,* 2004, p. 333.

[7]   Haas, R.W., and L. Zimmermann, "22GHz Measurements of Dielectric Constants and Loss Tangents of Castable Dielectrics at Room and Cryogenic Temperatures," *IEEE Trans. MTT,* No. 11, 1976, p. 881.

[8]   Carpi, F., and D. De Rossi, "Improvement of Electromechanical Actuating Performances of a Silicone Dielectric Elastomer by Dispersion of Titanium Dioxide Powder," *IEEE Trans. Diel. and Elect. Insul.,* Vol. 12, No. 4, 2005, p. 835.

[9]   TransTech Catalog, "Products for RF/Microwave Applications," 2003.

[10] Holtzman, E. L., "Wideband Measurement of the Dielectric Constant of an FR4 Substrate Using a Parallel-Coupled Microstrip Resonator," *IEEE MTT,* Vol. 54, No. 7, 2006, p. 3127.

[11] Nelco Materials Selection Guide, Rev. 7-10.

[12] Petrie, E. M., "Reinforced Polymers for High-Temperature Microwave Applications," *IEEE Trans. Elect. Insul.*, Vol. E1-5, No. 1, 1970, p. 19.

5

Low Dielectric Constant Ceramic Dielectrics

5.1   Introduction to Ceramic Dielectrics

There are an enormous number of possible ceramic dielectrics [1]. In practice, the range is constrained by the polarization of various types of cation-oxygen bonds. If we ignore nitrides and beryllia, the major dielectric types resolve around materials containing Si-O bonds and Al-O bonds for low dielectric constants; Ti-O, Zr-O, Ta-O, Nb-O, and W-O bonds for intermediate dielectric constants; and Ti-O bonds for high dielectric constants, essentially because of the increasing polarization of these bonds. Many of the compounds that include these groups are capable of intrinsically high Qs, but problems arise when we try to achieve temperature stability of the dielectric constant or very high dielectric constants relative to the intrinsic polarizability of the higher polarizable bonds, by influencing their crystallographic environment. There are also other molecular level contributions to dielectric constant, for example the presence of "lone pairs" of electrons in Bi^{+3} and Pb^{+2}, their influence recently redefined in [2]. This chapter will consider dielectrics with dielectric constants up to 20. The applications for such materials include substrates for microstrip circuits, patch antennas, matching transformers, coaxial and TM resonators, and supports for high dielectric constant TE resonators used as oscillator and filter elements.

Although the great majority of dielectrics in use today at microwave frequencies are oxides, increasing use is being made of nitrides and other nonoxides.

5.2 Measurement

The measurement of dielectric properties is discussed in Chapter 12. The main issue is the so-called QF rule. There is no intrinsic reason why a dielectric would follow such a rule over a wide frequency range, and many do not. Because the most reliable methods tend to use high-Q cavities, these are single frequency points and are seldom repeated over a series of frequencies under the same measurement conditions. However, the QF rule is a useful way of estimating changes over small frequencies changes, perhaps less than an octave, and a very rough indication, over wide, multioctave frequencies, of comparable performance.

When comparing very different methods, such as cavity perturbation or printed resonator methods, extreme caution must be used in comparing Q values. These methods are also not suitable for temperature coefficient (Tf) comparisons if extreme accuracy (< few ppm/°C) is required, and high-Q cavities are preferred. Comparing Tfs (Chapter 12) from different sources must also be viewed with caution, as the Tf of the same dielectric resonator will change significantly depending on the cavity size and bulk material used to manufacture it, regardless of its plated metal finish. Cavities of most common metals are roughly similar because their expansion coefficients are similar (Chapter 7), but silver plated Invar or very thin metalized plastic have zero and very large expansion coefficients, respectively, and will give very different results.

5.3 Applications

It is important to consider the application in choosing dielectrics. In many situations, such as coaxial or printed resonators or simple transmission lines, the losses are dominated by the metal of the transmission line and its proximity to its ground plane, or to the metal walls of the resonator. As a result, it is easy to over-specify the Q or dielectric loss contribution required from the material itself. Other nonresonant structures, like antennas, need relatively low Qs and minimal control of temperature stability of the dielectric constant. When considering resonant structures bounded by the dielectric itself, the issue can become radiation or unwanted coupling if the dielectric constant is too low, particularly at high microwave frequencies.

Applications have moved on in the last decade. Ceramic resonators and filters are not used in very high–volume applications like cell phones because bulk or surface acoustic wave devices (BAW and SAW) have better Qs for the same volume, particularly when the total height on a printed circuit board is considered. Coaxial ceramic resonator-based filters are still extremely useful for lower volume more specialized commercial and military applications. In cellular base stations, there has been a progressive move away from TE mode-based auto-tuned ceramic combiners to hybrid combiners or multichannel power amplifiers because of size or cost. TE mode filters using high-Q ceramics (typically with QF products > 40 K.GHz or 40 THz) are now largely confined to specialized applications like narrowband passband or notch filters where high Q is mandatory, particularly below 1 GHz where size, weight, and cost are issues. TM mode filters using ceramic do however represent a viable solution below 3 GHz for Qs in the 5- to 10-K region where they can compete with metal TEM mode filters for size and temperature stability (see Figure 5.1). Above about 2 GHz, TE mode filter ceramic resonators are sufficiently small and high Q to compete with TM mode for some applications where the small additional cost is not a factor.

From an applications point of view we will look progressively from the lowest dielectric constant ceramics to the highest. Although high dielectric constant and reduced Q are clearly associated over a wide range of types of materials, there are no absolute laws governing this relationship and attempts to generate rules have largely been empirical at best.

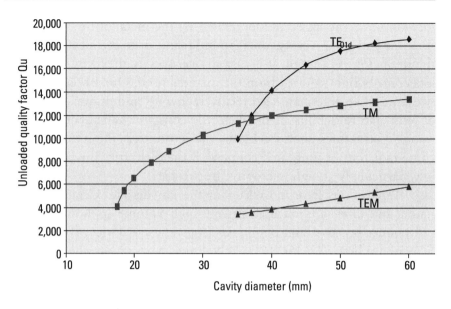

Figure 5.1 Q versus cavity diameter at 2 GHz for dielectric $TE_{01\delta}$, dielectric TM, and metal TEM modes.

5.4 Silica and Silicates

Silica, as quartz in various allotropic forms or as fused silica, has the lowest dielectric constant (3.78) of all dense ceramic dielectrics, because of the low polarizability of the Si-O bond. These forms also have very low Qs. The low dielectric constant makes fused silica very attractive as a support for higher dielectric constant resonators, as a tuner or tuner support, as the dielectric for high-frequency transmission lines, and so on. Its prime drawback is its very low thermal expansion coefficient of 0.7 part per million per degree Celsius (ppm/°C), which makes adhesion of metals in the 20 ppm/°C range difficult when deposited on fused silica. Gluing fused silica to other ceramics in the 10 ppm/°C range also has some issues, usually requiring very flexible adhesives. Some applications may justify the cost of single crystal quartz, which has highly anisotropic thermal expansion. Because of the BAW and SAW industries, such materials are available in relatively large sizes (300 mm or more). If the correct crystallographic cut is selected, expansion compatible with metal is possible. Note also that fused silica and natural quartz can be used as fillers in plastic/ceramic composites such as Fluorosint (Chapter 4).

5.4.1 The Range of Si-O–Based Dielectric Materials by Using Silicates

Silicate glasses used in composites were discussed in Chapter 4, so this chapter will only include crystalline materials (apart from fused quartz) and more specialized applications like low temperature cofired ceramic/glass composites (LTCC). The most useful of these silicates, from a microwave applications point of view, are Steatite and Forsterite based. The most useful compounds by polarizability of the various cation-oxygen bonds and hence dielectric constant are shown in Table 5.1, for the lower dielectric constants of interest.

There are a number of caveats to arranging these materials in this manner. Many are made directly from raw mineral materials and may contain significant amounts of other elements. For example, cordierite, which is not included here, may contain so much alumina that its dielectric constant rises to the same value as steatite or fosterite [3]. Other additives are introduced to improve firing or dielectric loss, also potentially affecting dielectric constant. Although it may seem logical to blend such materials to give a range of dielectric materials, the dynamics of commensurate ceramic phases in phase diagrams apply, and some combinations may produce completely different phases and hence subtly different dielectric constants and loss. Phases are selected to *not* react with each other (such as ilmenite and perovskite phases respectively in magnesium titanate/calcium titanate) where mixtures of attributes are wanted. In other situations, a complete solid solution range such as the magnesium aluminum titanates in Table 5.1 may be desired.

Systems containing Ti^{+4} in particular are susceptible to reduction to Ti^{+3}, creating loss through electron hopping between the two valencies, and some chemistries may favor Ti^{+3} formation, particularly when Al^{+3} is present. In general, most commercially available materials in this range have losses in the 0.0002 to 0.0005 range, with the exception of quartz/fused silica and alumina, which can be significantly lower. A low dielectric constant (~4.5) alternative to fused silica is cordierite ($Mg_2Al_4Si_5O_{18}$), which is not a glass and can be made by convention-

Table 5.1
Common Magnesium-Based Ceramics for Microwave Applications

Chemistry	Contributing Bond Polarizations	Crystallographic Systems	Microwave Dielectric Constant Range
SiO_2	Si-O	Quartz fused silica (Glass)	3.8
$MgSiO_3$	Si-O, Mg-O	Steatite	4–6
Mg_2SiO_4	Si-O, Mg-O	Forsterite	6–7
Al_2O_{3-}	Al-O	Alumina	9.5–9.9
$(Mg,Al,Ti)O_5$	Mg-O, Al-O, Ti-O	Pseudo-Brookite	9–14
Mg_2TiO_4	Mg-O, Ti-O	Inverse spinel	13
$MgTiO_4$	Mg-O, Ti-O	Ilmenite	16

al ceramic processing. It has slightly higher loss and greater thermal expansion than fused silica.

Some generalizations are possible, however. Purer grades of starting materials generally produce lower dielectric loss. Mg silicates made from relatively impure starting materials may have a dielectric loss of 0.0002 at X-band. This can be improved by a factor of 4 or more with pure materials. For example, forsterite (Mg_2SiO_4) has been made in purified form with a Qf product of 240,000 and a dielectric constant of 6.8. Additions of 1% TiO_2 have achieved similar Qf but with a Tf of -65 ppm/°C (TE of +90 ppm/°C).

In general, we would expect material made exclusively from elements with invariant valency like Mg^{+2} and Si^{+4} to achieve the lowest losses, although Mg^{+2} materials can be susceptible to moisture or carbon dioxide if MgO is present, and MgO itself is not usable as a standalone material for that reason for microwave applications.

Processing of Ti^{+4} containing materials is subject to both impurity and high temperature–related reduction. Introduction of impurities can come from starting materials, binders, milling media, lubricants and crosscontamination with other materials being processed. Reduction at high temperature can be avoided by binder burn-out control, oxidizing atmospheres, the use of lower firing temperature additives, or the addition of oxidizing manganese oxides.

There are Zn-based analogs for most of the Mg systems, including Willemite (Zn_2SiO_4) and $ZnTiO_3$, which can be made temperature stable by suitable additions. Currently, however, Mg-based systems are preferred commercially.

Materials using combinations of calcium titanate and magnesium titanate can be used to produce temperature-stable materials close to a dielectric constant of 20 (Chapter 6).

5.5 High-Temperature and High-Conductivity Materials

5.5.1 Nitrides, Oxides, and Fluorides

Ceramic nitrides can be compared with more traditional oxide-based ceramics on the basis of their chemistry. N^{-3} has the same extremely stable electronic structure as O^{-2} and F^{-1}, $1s^2\ 2s^2 2p^6$, and therefore tends to form the same type of strongly covalent bonds (energy levels in an atom are indicated by s, p, d, f, and so on, with the maximum for s being two paired electrons and six paired electrons for p levels, so in these ions each level is full, implying stability). Three nitrides that have become widely used are the graphite like structured boron nitride (BN), the hexagonal beta-phase silicon nitride, (Si_3N_4), and the wurtzite-structured aluminum nitride (AlN). The criteria for high thermal conductivity in the context of oxides and nitrides are thought to be low atomic mass, highly directional covalent bonding, highly ordered crystalline structure, high purity, and low crystal defects. These criteria are shared with silica and boron oxide except in one regard. Silica and boron oxide structures allow flexibility of covalent bond orientation and so form amorphous glassy materials, an effect that greatly reduces thermal conductivity. Single crystal quartz is also a poor conductor, indicating the problem lies in the relatively large, easily displaced SiO_2 structure. Other than low atomic mass, these criteria are also useful in predicting low dielectric loss in oxides where the cation has a stable valency, and to a degree, low dielectric constant. By extension, then, nitrides meeting these critera should also show high thermal conductivity and low dielectric loss, which they do with as materials with increasing purity and reduced defects have evolved. Silicon nitride is discussed further in Chapter 10, but Table 5.2 shows its similarity to boron nitride and aluminium nitride (in particular its thermal stability). Their dielectric properties appear in more detail in Tables 5.4 and 10.3, as they are very dependent on the effect of sintering aids. Their high-temperature stability in inert atmospheres, low pressure, or vacuum makes them particularly useful for specialized industrial, space, high altitude, and microwave tube applications.

Some fluorides are used in combined microwave and ultraviolet to infrared applications such as radomes for radar and infrared measurement systems, and cryogenic applications. Their dielectric, thermal, and light transmission characteristics are shown in Table 5.3.

The temperature range of fluorides is limited by sensitivity to moisture which affects both dielectric loss and light transmission [4]. They have reasonably good thermal conductivity, but relatively modest relative to the best oxides and nitrides. They have low dielectric loss, particularly at cryogenic temperatures, but are not temperature stable dielectrically [5]. Lithium fluoride is weak mechanically so is not used much.

5.5.2 Alumina (Al_2O_3)

Alumina is a good example of a material where the application should suit the grade of material used. For microstrip applications where the

Table 5.2
High-Temperature Properties of Nitride Dielectrics

Dielectric	Structure	Maximum Use Temperature in Air Atmosphere	Maximum Use Temperature Inert Atmosphere	Approximate Dielectric Constant	Approximate Dielectric Loss
BN	Graphite-like	1,000°C	2,800°C	4	< 0.0005
AlN	Wurtzite	Some oxidation at 700°C	1,800°C	9	< 0.001
Si_3N_4	Hexagonal beta-phase	1,850°C decomposes	1,850°C decomposes	8	0.002

Table 5.3
Dielectric, Thermal, and Light Transmission Properties of Fluorides

Dielectric	Crystal Class	Dielectric Constant	Dielectric Loss at 16.5 GHz	Thermal Conductivity W/mK	Melting Point/ Maximum Use (°C)	Light Transmission Range, micron
MgF_2	Tetragonal	5.165 at 32.9 GHz	0.00008	14–15	MP 1585 use 500	0.1 to 7.5
CaF_2	Cubic	6.844 at 17.5 GHz	0.0002	10	MP 1425 use 1,000	0.1 to 7
LiF	Cubic	8.90 at 16.5 GHz	0.0002	11.3	MP 870 use 600	0.1 to 6

effective Q may be a few hundred, a material Q of 10 times that figure may be adequate. In the range 1 to 10 GHz, this implies QF products in the 1,000- to 10,000-THz range, achievable with relatively low grades of alumina (see Table 5.2). At the other extreme, a resonator may be required to achieve a QF product in the range 50,000 to 200,000 THz, implying similar values for an alumina support. Although this may imply purities in excess of 99.9% for the latter application, it cannot be assumed that the purity itself is the only consideration. Very small additions of TiO_2 have been shown to improve QF product [6], and larger amounts as nano-sized additions also improved the temperature coefficient, Tf [7]. Microwave grades of alumina are used extensively as high-Q resonator supports. A general guide to the properties of alumina can be found in the Selected Bibliography for this chapter.

Other materials systems with dielectric constants below 20 include other oxides and some nonoxides (mainly the nitrides) although perhaps surprisingly include diamond (but not other forms of carbon).

5.5.3   Boron Nitride (BN)

Boron nitride can be made [8] in the form of an anisotropic, machinable dielectric. Its main drawback in that form is that it is too soft and mechanically weak to be used in most applications. It also has a very low tribo-electric coefficient, like Teflon, and is thus difficult to glue. However, it has low dielectric constant and loss at microwave frequencies, has high thermal conductivity, and will withstand high temperatures in vacuo. Because it is hot pressed, its graphite-like structure will align with its plane perpendicular to the direction of pressing, giving it anisotropic mechanical, thermal, and dielectric properties. Most commercial grades contain boron oxide, which combined with slight porosity cause the material to absorb water vapor, making it high loss dielectrically under humid conditions. This can be avoided with grades containing calcium borate or no additives, but in any case is not necessarily a problem in a vacuum. For these reasons and its high relative cost, boron nitride is used primarily in microwave vacuum tubes. There are chemical vapor deposition forms that are highly anisotropic, but these are used rarely because of cost. Highly oriented pyrolytically deposited films of BN have extreme ranges of thermal properties because of the degree of orientation, but are not used commercially.

5.5.4 Beryllium Oxide (BeO)

BeO is potentially a very useful ceramic for microwave support structures and heat sinks [9]. However, the toxicity of BeO dust makes it difficult to manufacture, machine, and subsequently handle. It has low dielectric loss and a low dielectric constant, making it a candidate for resonator supports, supports for the metal structures in travelling and backward wave tubes, and heat sinks for high-power semiconductor devices, and is still often used for these applications, where its thermal conductivity and high strength are also major assets. BeO is normally manufactured and machined in specialized facilities because of its toxicity, making it a relatively expensive material. Currently, only aluminum nitride is a reasonable alternative for most relevant applications.

5.5.5 Aluminum Nitride (AlN)

AlN is made [8, 10] by hot pressing typically, making it potentially expensive, but is not anisotropic. It has much higher thermal conductivity than alumina, but to date has not achieved the extreme QF product of alumina. As a result, its main use has been as a microwave heat sink or microstrip substrate where a lower Q is tolerable, displacing alumina for many applications. As purer and more defect-free materials become available, the dielectric loss has fallen, and wider ranges of applications have opened up. In inert atmospheres it can operate up to 1,800°C, higher than alumina.

5.5.6 Diamond

Synthetic diamond is now widely available commercially. Although typically seen as a semiconductor, intrinsic diamond is an insulator. The dielectric constant has not been reported accurately so far, but can be estimated by the square of the optical refractive index, giving a minimum figure of 5.76. Its main advantage is its remarkable thermal conductivity, around 10 times the value of the best oxides or nitrides. Although expensive, diamond heat sinks for microwave devices look very useful for high power or very densely populated device environments.

The properties of candidate support materials are summarized in Table 5.4. To calculate the QF product in THz, the inverse of the loss tangent is multiplied by the frequency in gigahertz.

5.6 Dielectrics for Thick Film and Low Temperature Cofired Ceramic (LTCC) Applications

Most of the oxide ceramics discussed so far have firing temperatures in the 1,200–1,600°C range. For thick film and LTCC, firing temperatures have to be in the range of 850–900°C typically. For LTCC transmission lines and thick-film "crossover" applications, low dielectric constants are used. A further constraint is matching expansion coefficients of metal, dielectric, and dielectric substrates, where the substrate may be alumina or aluminum nitride, for example. Dielectric losses are determined by the required Q of the transmission line and are relatively modest, but temperature coefficients of dielectric constant also matter for two- or three-dimensional filter elements, for example. Most thick-film or LTCC materials are glass/ceramic composites used to achieve low firing temperatures. The glass base is usually silicate, borosilicate, or oxide/borate. Environmental restrictions now limit the use of PbO-based glasses. The ceramic dielectric is typically Al_2O_3, TiO_2, titanate, or zirconate based to raise the dielectric constant. Typically the glass limits the range of dielectric constant and dielectric loss. Typical values used are around 6 for transmission lines and circuit elements, with effective Qs greater than 300. For very high dielectric resonators or some magnetic circuit elements, where low firing temperatures are not possible, the tendency is to embed these within the LTCC structure as

Table 5.4
Low Dielectric Constant Microwave Materials for Device Supports and Heat Sinks

Material	Dielectric Constant (X-band)	Dielectric Loss/ Frequency (GHz)	Thermal Conductivity (W/m.K)	Thermal Expansion (ppm/C, RT)	Temperature Coefficient of E (T_E, ppm/°C)
Cordierite	4.5–6	0.0002/9.3	3	2.0	+55
Steatite	5.6	0.005/9.2	2–3	7.5	
Forsterite	6.3–6.8	0.0002/9.3	3	10.0	+107
Alumina 99.5%	9.6	0.0001/10	33.6	6.0	+94
Alumina 94%	8.8	0.0009/10	19.3	6.2	
Beryllium oxide	6.7	0.0009/10	256	8.0	
Boron nitride (B_2O_3 additive, perpendicular)	4.6	0.0017/8.8	30	11.9	
Boron nitride (B_2O_3 additive, perpendicular)	4.2	0.0005/8.8	33	3.1	
Boron nitride no additive, perpendicular	4.0	0.0012/8.8	71	11	
Boron nitride perpendicular	4.0	0.0003/8.8	121	4.5	
Aluminum Nitride	8.9	0.001/10	170	4.5	
Diamond	5.76	~0.001	2,000	1.2	

discrete prefired elements. Some typical thick-film LTCC film tapes for microwave applications are listed in Table 5.5.

A recent development has been the use of nano-metal conductors for thick-film applications with firing temperatures around 300°C. This in turn has led to the need for ultra-low firing dielectric compositions. These, if practical, would allow circuit deposition on to high-temperature plastics and ceramic/plastic composite materials or even high-temperature plastic fiber–based clothing for wearable applications like antennas. Currently this is only practical with expensive thin-film or electro-deposition techniques.

Table 5.5
Microwave and Thermal Properties of LTCC Tapes

Product	Chemical System	Dielectric Constant	Dielectric Loss	Expansion Coefficient	Firing Conditions
Ferro A6-S	Ca borosilicate	5.8 (1–90 GHz)	0.002 (1–90 GHz)	> 8 ppm/°C	850°C
Dupont 9K7	Al_2O_3 glass	7.1 (9–10 GHz)	0.001 (9–10 GHz)	4.4 ppm/°C	850°C
Heraeus CT707	Lead free	6.3 (2.5 GHz)	0.003 (2.5 GHz)	7.6 ppm/°C	850°C

5.7 Summary

Low dielectric constant materials are useful for many support or heat sinking applications, but also are used for high temperature applications, where oxides have been partly replaced by nitrides. Fluorides have some unique microwave and optical characteristics for combined applications like radomes. There are certain common materials characteristics between materials that have low loss and high thermal conductivity that are being exploited as purer and more defect-free materials are manufactured.

References

[1] Narang, S. B., and S. Bahel, "Low Loss Dielectric Ceramics for Microwave Applications: A Review," *J. Ceram. Proc. Research*, Vol. 11, No. 3, 2010, pp. 316–321.

[2] Payne, D. J., et al., "Electronic Origins of Structural Distortions in Post-Transition Metal Oxides," *Phys. Rev. Lett.,* Vol. 96, 2006, pp. 157–403.

[3] Trans Tech catalog, *Products for RF/Microwave Applications*, pp. 1–30, http://www.trans-techinc.com.

[4] www.corning.com.

[5] Jacob, M. V., "Low Temperature Microwave Characterization of Lithium Fluoride at Different Frequencies," *Sci. Tech. Adv. Mat.*, Vol. 6, No. 8, 2005, pp. 944–949.

[6] Alford, N. M., and S. J. Penn, "Sintered Alumina with Low Dielectric Loss," *J. Appl. Phys.*, Vol. 80, 1996, pp. 5896–5898.

[7]   Huang, C. L., et al., "Microwave Dielectric Properties of Sintered Alumina Using Nano-Scaled Powders of Alumina and TiO_2," *J. Amer. Ceram. Soc.*, Vol. 90, 2007, pp. 1487–1493.

[8]   www.accuratus.com.

[9]   www.americanberyllia.com.

[10] www.ceradyne.com.

Selected Bibliography

Gitzen, W. H., "Alumina as a Ceramic Material," *American Ceramic Society Special Publication*, No. 4, 1970.

6

High Dielectric Constant Dielectrics

6.1 Introduction

At low dielectric constants, it was possible to find very simple oxide systems with one or two cations. The simplest systems are SiO_2 and Al_2O_3. Both exist in single crystal form, as quartz and sapphire, respectively, and we find high-purity polycrystalline versions of fused silica and alumina, which approach single crystal values of Q. Binary compounds of these and MgO achieve QF products well in excess of 100 THz. Single crystal rutile illustrates some of the problems of higher dielectric constant materials. TiO_2 is subject to reduction to Ti^{+3} because of atmospheric reduction during high temperature processing, and because it can easily be reduced by other cations by redox reactions. However, single crystal QFs for TiO_2 have been reported of 1,000 THz at 77K, suggesting that is not the issue fundamentally, although paramagnetic impurities appear to play a part at these temperatures. TiO_2 has a high dielectric constant, about 100, but a very high negative T_E of about −600 ppm/°C, making it impractical for virtually all applications. In rutile, its preferred form, the crystal structure is tetragonal consisting of parallel chains of octrahedraly coordinated Ti^{+4} surrounded by 6 oxygen ions. The 6 oxygen-titanium bond lengths are equal in the octahedron, and we can assume the dielectric constant and its T_E is primarily due to the polarization of these bonds and how they change with temperature.

The octahedrons share sides such that there are voids in the structure. The problem can be regarded as two-fold. Can the voids be manipulated such that the octahedrons occupy different positions relative to each other (tilting), and can other cation-oxygen bonds present be adjusted to change the dielectric constant and T_E? Secondly, can the Ti-O bond itself be altered by changing the interatomic distance, to change its polarization and hence dielectric properties, particularly if not all bonds are changed equally such that the Ti^{+4} ion is "off center"? The same type of arguments can be made for multication structures containing Nb-O, Ta-O, and Zr-O bonds, for example, for lower ranges of dielectric constants. We will consider a range of compounds and the likely loss mechanisms that dominate in each. One generalization that can be made, however, is that in multication, multisite compounds, the number of different cations and sites require an increasing degree of ordering, that is the need for each cation to be in the correct position in the structure, to achieve very high–QF products, although there are some exceptions to this general rule.

A third consideration for polycrystalline materials is second phase material, which typically gathers at the grain boundaries. This relates to control of stoichiometry. In garnets like YIG, compositions are made slightly iron deficient because second-phase yttrium orthoferrite ($YFeO_3$) is relatively harmless in terms of affecting magnetic and dielectric properties. For similar reasons, dielectric compositions are often deliberately biased towards a preferred second phase to avoid another unwanted phase. Nevertheless, the presence of a second phase will not give ideal dielectric properties; it may lower the Q slightly or cause an unpredictable temperature coefficient if the amount is not tightly controlled. The boundary between primary and second phases may be highly disordered, which in itself may reduce Q, even if both phases are themselves high Q.

6.2 Dielectrics with Dielectric Constants in the Range 20 to 55

Oxide systems that are reasonably high Q and have a stable temperature coefficient abound. We will discuss the most useful of these in detail. These can be classified into broad types, those with high-QF products, which are temperature stable, those with useful QF products, which are reasonably temperature stable, and those which are not temperature stable but are useful for many microwave applications. In general, QF products quoted in this chapter are measured at 1 to 3 GHz as $TE_{01\delta}$ resonators, where the resonator diameter is more than 1/3 of the cavity diameter. Values quoted in the literature are often measured at 10 GHz, often at unspecified resonator/cavity ratios. The latter will tend to give much higher QF values, which will not result in proportional values at lower frequencies in practical cavity sizes. When selecting materials, it is important to find data that takes these factors into account. Although it can be argued that large cavity/resonator ratios reflect the absolute loss tangent, experience has shown that this is often not the case in the range 1 to 3 GHz, where most very high–Q requirements are encountered.

If we look at the last category first, the most useful one is the Mg-TiO_3/$CaTiO_3$ series, which will produce a dielectric constant range well in excess of 55 (> 150 at 100% $CaTiO_3$). It depends on a two-phase mixture of positive T_E ($MgTiO_3$) and strongly negative T_E ($CaTiO_3$), and includes a compensation point where T_E is 0 and a slightly different one where Tf is 0. The variation of dielectric constant versus composition is shown in Figure 6.1. A typical dielectric loss of materials from this system for modest Q applications is 0.0015 at 9 GHz.

The tem perature-stable region, when adjusted to produce Tf of close to 0 ppm/°C, is typically modified in composition to produce higher QFs, around 20 THz. The system does suffer from drawbacks because it depends on very small amounts of $CaTiO_3$ to compensate at 0 Tf. This means essentially that very tight control of the compensation point is only possible if the Mg/Ca ratio is exactly correct. The Tf, in common with other systems of this type, if set to 0 at room temperature, will not be 0 at other temperatures, producing a nonlinear Tf, expressed as T'f in ppm/°C/°C, which in this case at room temperature is −0.09 ppm/°C/°C. This means that over 11°C, there will be a 1 ppm change in the measured Tf, making the material unsuitable for very high-Q narrowband filters, but acceptable (Figure 6.2) for lower-Q wideband devices where frequency drift is not a major concern.

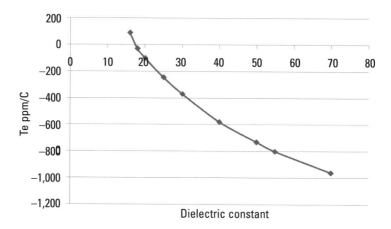

Figure 6.1 MgTiO$_3$/CaTiO$_3$ system-dielectric constant (X-axis) versus T$_E$, in ppm/°C.

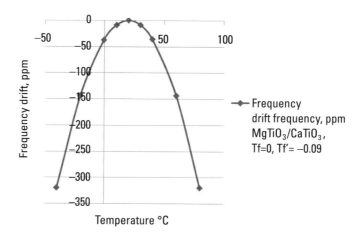

Figure 6.2 Effect of nonlinearity of Tf on frequency drift, MHz, versus temperature (X-axis) for the MgTiO$_3$/CaTiO$_3$ system.

6.3 The BaTi$_4$O$_9$/Ba$_2$Ti$_9$O$_{20}$ System

This system produces dielectric constants of around 38 and QFs of about 20 THz at 3 GHz. The temperature coefficient depends on the tetratitanate/nonatitanate ratio but can be adjusted to be close to 0. This system is used extensively for microwave dielectric substrates, including tape cast products, and coaxial resonators where it can be hard die or extruded, less so for high-Q $TE_{01\delta}$ resonators. Systems based on this composition have been substituted with Zn and small amounts of Nb and Ta to raise the QF to ~40 THz at 2 GHz, creating additional Ba, Ti, Zn phases and a very complex overall phase structure [1]. These systems have exceptional Tf linearity, and the sign of the T'f can be modified by variation of the tetratitanate and nonatitanate ratio

6.4 The Zirconium Titanate/Zirconium Tin Titanate System $(ZrTiO_4/(Zr,Sn)TiO_4)$

This has similar dielectric constants (36) and QFs to the BaTi system. When Zr^{+4} is substituted with Sn^{+4}, or Zn^{+2} and Nb^{+5}, higher Qs can be obtained, in the region of 40 THz, and also higher dielectric constants (up to ~46). Dielectrics from both ends of this range can have similar applications, particularly substrates and TE resonators. The structure is orthorhombic alpha-PbO.

6.5 Perovskite Materials

Perovskites cover a wide range of dielectric constants with Tfs close to 0 ppm, typically 20 to ~55. They can be characterized by titanates with QF in the 40- to 50-THz region, and niobates, tantalates, and tungstates in the 80- to 110-THz range at 2 GHz or more (as mentioned, QF is of course not necessarily a constant, and comparison of values at 2 GHz and 10 GHz may give very widely different values, with 10-GHz values typically higher by up to a factor of two for perovskites of this type). In general, QFs of the niobates, tantalates, and tungstates are inversely proportional to the dielectric constant, which may be a function simply of the polarizability of the Nb-O, Ta-O, and W-O bonds (see Section 6.6). Titanates with the same dielectric constants, however, have significantly lower QF.

Useful perovskite titanates can be created by $CaTiO_3$ (structure shown in Figure 6.3), with a dielectric constant of about 170 and Tf of +800 ppm/°C, combined in solid solution with rare earth aluminate perovskites like $LaAlO_3$, with a dielectric constant about 20, and

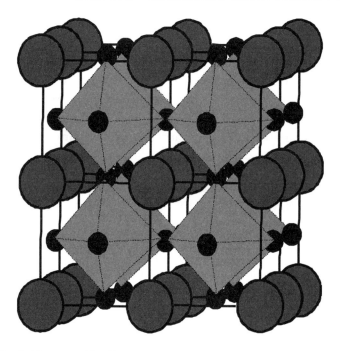

Figure 6.3   Cubic perovskite (large spheres are A-site atoms, the small spheres are oxygen ions).

a small negative Tf, to give a Tf close to 0 ppm. Note that in Figure 6.3, the A-site (Ca or La) ions are drawn as the larger spheres, and the oxygen ions are the smaller spheres, with the titanium or aluminium B-site ions inside the octahedra. In reality, of course, the oxygen ions are much bigger than the cations. Goldschimdt [2] used a tolerance factor to estimate the stability of perovskite structures based on ionic radii of oxygen and of the A and B site cations in ABO_3, and predicted octahedral tilting of the TiO_6 octahedra was the outcome of variation of the tolerance factor $Ra+Ro/2^{1/2}(Rb+Ro)$, where Ra, Rb, and Ro are the respective ionic radii of the A, B, and oxygen ions. The unity tolerance factor indicates a cubic perovskite; significant tilting is thought to occur at a ratio of around 0.81, creating a shift from cubic to orthorhombic symmetry, making some assumptions about "undersized" A-site cations. Stability is thought to lie in the range 0.78 to 1.05, although this assumes completely ionic bonding and the absence of Jahn-Teller ions. Octahedral tilting is thought to affect both Tf and T'f in dielectrics, and normally tolerance factors close to unity are desirable, limiting the many combinations of A and B site ions that are possible. Combinations of 70% molar $CaTiO_3$ and 30% molar $NdAlO_3$ to form a single phase rhombic perovskite give a dielectric constant of 43.5 and Tf − 2.1 ppm [3]. Analogous to the titanates are the calcium barium zirconates. These have useful dielectric constants (~30) and QF's,(20 THz) with stable Tfs, but have not found universal use [4].

Theoretical calculations of the dielectric constant of perovskites have been made on the basis of ion polarizability from their Born radii, and the dielectric loss on the basis of vibrational modes at calculated phonon frequencies. These modes are influenced by the tolerance factor, and by disorder, suggesting that dielectric loss, when all other contributions are omitted, is ultimately determined by these factors at least in perovskites of this general type [5]. Actual measurements of such vibrational modes by infrared and Raman spectroscopy on similar perovskites have shown agreement with this type of approach.

A summary of useful commercially available ceramic dielectric with dielectric constants in the range 20 to 55 and moderate Qs are given in Table 6.1. Extreme caution should be used in comparing QF products at widely different frequencies.

Table 6.1
Moderate-Q Dielectrics in the Range 20 to 44

Chemistry	Dielectric Constant	Temperature Coefficient, Tf	QF Product	Source
MCT	20–21	–5 to +5 ppm/°C	70 THz at 12 GHz	TransTech, NTK, Morgan, Kyocera
BT_4/B_2T_9	36–38	+6 ppm/°C	20 THz at 3 GHz	TransTech, PicoFarad
Zr, Sn Titanate	36	–4 to +15 ppm/°C	42 THz at 2 GHz	Morgan, NTK
BT_4/B_2T_9 +Zn, Nb	35–36.5	–3 to +9 ppm/C	45 THz at 2 GHz	TransTech
$CaTiO_3/Ln/Mg/Ti$	39–42	–3 to +8 ppm/°C	65 THz at 4 GHz	Kyocera
ZrTitanate + Zn, Nb	43	–3 to +9 ppm/°C	45 THz at 2 GHz	TransTech, NTK, Morgan
$CaTiO_3/LnAlO_3$	44	–7 to +8 ppm/°C	45 THz at 2 GHz	Kyocera

6.6 High-Q Perovskites

These are based on niobates, tantalates, and tungstates. Although they have appreciably higher QF products than titanate perovskites, much of the above applies. The main difference lies in the polarizability of the B-site cation-oxygen bond, which is significantly lower in these perovskites, and which results in lower dielectric constants, niobates being generally highest, and tungstates lowest. This in turn may lead to higher Qs (lower dielectric losses). The niobates and tantalates are called 1:2 ordered perovskites because the B site typically contains 1 divalent ion (B' is typically Sr, Ca, Mg, or a first transition element) for each 2 pentavalent ions (B" is Nb or Ta). The A-site contains Ba, Ca, or Sr. Combinations of A-site La^{+3} and B'-site Li^{+1} can be partially substituted [6]. The vibrational modes and infrared and Raman spectra of BaB'B" perovskites have been extensively reported, particularly where B' is Mg or Zn, and ordering directly observed using high-resolution electron microscopy (HRTEM). Because of the high cost of tantalum ores, only $BaZn_{1/3(1-x)}Co_{1/3x}Nb_{2/3}O_3$ (BZCN) is used at low microwave frequencies (2 to 5 GHz) and $BaMg_{1/3}Ta_{2/3}O_3$ (BMT) and $BaZn_{1/3}Ta_{2/3}O_3$ (BZT) at high frequencies (> 10 GHz, where small-sized resonators are used). A related perovskite is $BaMg_{1/2}W_{1/2}O_3$ (BMW). The infrared and Raman spectrum of the latter was first published by Blasse [7]. It has a dielectric constant of 20, a Tf of -30 ppm/°C and a QF product of 130 THz, and can be modified using Ta to a 0 ppm composition [8]. All of the high-Q perovskites present difficulties in manufacture because of the degree of B-site ordering required to attain maximum Q values. Zinc volatility and multivalent cobalt add complexity to processing, as does the need to attain precise stoichiometry with high purity and expensive oxides. BZCN is often made nonstoichiometric to form the high-Q 8:1:6 hexagonal perovskite as a second phase [9]. Attempts to use a Co-free version of BZCN by using Sr-substituted BZN result in driving its positive Tf higher and more nonlinear (Figure 6.4 and [10]), which may be the result of octahedral tilting. As a result, these materials can only be used where the additional performance justifies the extra cost. Their properties are summarized in Table 6.2.

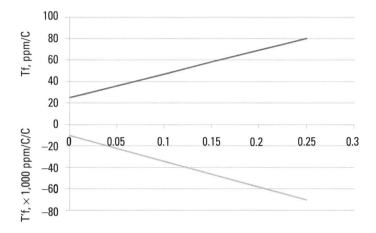

Figure 6.4 Effect of Sr substitution, x in $Sr_xBa_{(1-x)}Zn_{1/3}Nb_{2/3}O_3$ (X-axis) on Tf and T'f [10].

Table 6.2
High-Q Perovskite Dielectrics

Chemistry	Dielectric Constant	Temperature Coefficient, Tf	QF Product	Source
BMW	21	−2 to +4	150 THz at 5 GHz	Kyocera
BMT	24	−2	200 THz at 10 GHz	NTK
BZT	29	+5 to −5	110 THz at 2 GHz	TransTech, NTK, Morgan
BZCN	33–35	+5 to −5	90 THz at 2 GHz	TransTech, TDK, Morgan

6.7 Temperature-Stable Dielectrics with Dielectric Constants Greater Than 55

Relatively low loss materials like rutile (TiO_2) and perovskite $CaTiO_3$ both have very high dielectric constants but very large positive Tfs. Despite intensive efforts, no temperature-stable perovskites have been found in this range of dielectric constants, and the nearest simple compound has been (Pb,Ca)ZrO$_3$ [11] with Qs of only 1 to 3,000, but dielectric constants of 100 to 118.

The only system that has looked likely to approach good Qs with temperature-stable high dielectric constants has the tetragonal tungsten bronze (TTB) structure. The formula unit of the basic composition can be considered as solid solution series $A^{+2}_{(6-3x)}Ln^{+3}_{(8+2x)}Ti^{+4}_{18}O_{54}$; the most effective compositions lie between x = 0.5, and x = 2/3, the former corresponding to $A^{+2}_{(4.5)}Ln^{+3}_{(9)}Ti^{+4}_{18}Q_{54}$, or "114" as A:Ln$_2$:4Ti. For most work, A has been Ba^{+2}, but Pb^{+2} has been used. However, Pb^{+2} (and Bi^{+3} in the B-site) both reduce Q substantially while increasing the dielectric constant. This has been attributed to "lone pairs" of electrons from hybridization of the Pb^{+2} ions 6s and 6p energy levels (Bi^{+3} has a similar such pair) distorting the crystal environment and increasing the resultant polarization of the Pb-O bond, and hence (indirectly) the dielectric constant. Combinations of both A-site Pb^{+2} and B-site Bi^{+3} have produced materials with dielectric constants of 110, but QF products of only 500 to 2,500, depending on the concentration of each. A further issue is that Pb (and potentially Bi) containing TTBs have anomalous dielectric losses at low temperature, around −20°C, and poor nonlinearity at power, leading to potential harmonic distortion and intermodulation issues in microwave filter applications. This effect may be a form of incipient ferroelectric or paraelectric (Chapter 11) behavior. Interestingly, the model of isolated lone pair electrons has been revised (Chapter 5, [2]) to include involvement of the electrons from oxygen's 2s electrons in metal oxides of Pb and Bi, which may have implications for magnetic ferroelectric (multiferroic) materials.

When Ln A-site ions are Sm or Nd, typically when x < 0.5, the dielectric losses rise, and when x > 0.5 the dielectric constant begins to fall, although the QF product improves, and may peak at x = 2/3, which has been suggested as the lowest strain structure. As the Ln^{+3} ion becomes smaller, the dielectric constant falls but the QF product improves, with Sm^{+3} giving the best results in terms of a balance between Q and Tf in a series from La^{+3} to Sm^{+3} (Figure 6.5). The TTB structure is essentially planar, with a unit cell of 12.2 × 22.3 × 3.85 Angstroms (1A = 10^{-8} cm), a volume of about 1,047 A^3, although some publications use 7.7A, that is × 2, for the c dimension. Figure 6.5 shows the expansion of the unit cell in cubic Angstroms with Ln^{+3} ion substitution in Ba$_{4.5}$Ln$_9$Ti$_{18}$O$_{54}$, for the lanthanide series Eu (1,036

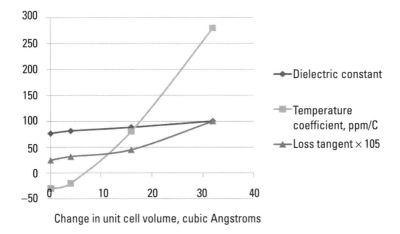

Figure 6.5 Change in unit cell volume (X-axis, cubic Angstroms) versus dielectric constant, temperature coefficient, and loss tangent (x 10^5) for $Ba_{4.5}Ln_9Ti_{18}O_{54}$.

A^3), Sm, Nd, and La (1,068 A^3) and the effect of the expansion on dielectric constant, temperature coefficient, and loss tangent ($\times 10^5$). The Gd^{+3} analog is difficult to synthesize, and smaller rare earth analogs do not form at all. Temperature-stable compositions are found with roughly 80% Sm and 20% Nd, or 90% Sm and 10% La at x = 0.6, and with QF products of 10 THz at about 3 GHz [12, 13]. Intensive efforts continue worldwide to raise this QF product without significant reduction of dielectric constant, including substituting for Ti^{+4} with similar sized ions, as suggested by Shannon [14] and Prewitt's tables of ion size and coordination.

In the TTB structure, in the XY plane in the extended unit cell, there are interconnected blocks of nine perovskite-like corner-sharing TiO_6 octahedra. For x = 2/3, the partly filled interstitial sites are 4 Sm^{+3} (A1) ions in the blocks in cuboctahedron 12-coordination, and Ba^{+2} ions (A2) in pentagonal prism 15-coordination interstices (Figure 6.6). In the z-plane, octahedral tilting of the TiO^6 octahedra is observed [15]. Ohsato also considered that higher dielectric constants were caused by enlargement of A-site polyhedra, depending on the Ba/Sm ratio (from the larger, 1:2 (114) to the smaller 4:9.33).

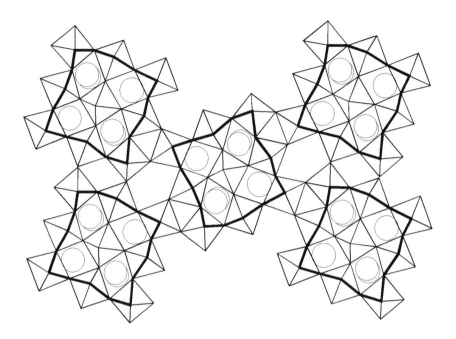

Figure 6.6 Part of the TTB structure. Note that (A1) ions are in quadrilateral interstices and Ba (A2) ions are in the pentagonal interstices.

6.8 Commercially Available TTBs

Commercial materials contain TTBs with Pb or Bi to raise the dielectric constant to 90 or more, typically with an increasing penalty in Q. However, the applications for these materials are normally coaxial resonators or substrates with microstrip transmission lines where Q and TF are less important than size reduction. Control of Tf is more difficult with TTs because the manufacturing method may orientate the plate-like structure. For example, die pressing will orientate the plates perpendicularly to the pressing direction, while extrusion will align them with the extrusion direction. Different orientations will give different Tfs and dielectric constants.

TTBs have essentially nonlinear Tfs, perhaps because of octahedral tilting, and certainly because many commercial materials contain second phases. Added to the orientation effect, it is usually better not to use TTBs for filters with very tight temperature drifts. It is unlikely any high dielectric constant TTB, where Er > 85, is capable of better

Table 6.4
Commercially Available Tetragonal Tungsten Bronze-Based Dielectrics

Chemistry of TTB	Dielectric Constant	QF Product	Source
Ba,Sm,Ti	75	10,000	TransTech
Bi,Ba,Nd,Ti	90	4,500	TransTech
Ba,Sm,Ti	76.5	10,000	Morgan
Ba,Nd,Ti	88	5,500	Morgan
	82	8,000	Maruwa
	110	3,300	Maruwa
	78	9,900	NTK
	93	4,800	NTK
	68.5	13,000	Kyocera

than +/– 1 ppm over more than very short temperature ranges, most are significantly worse. Only Morgan reports nonlinear properties over wide temperature ranges (Table 6.4). For their dielectric constant of 76.5, they report 0.02 ppm/C/C, and for their 88 material, 0.12 ppm/C/C over –30°C to +85°C. The effect of such a level of nonlinearity is seen in Figure 6.2.

The correlation between dielectric constant and QF product can be seen from the table. Research continues to try to improve beyond current commercial values, but the indication is that it is unlikely present boundaries will be exceeded by more than 20–30% on QF product for a given dielectric constant.

For higher dielectric constants, perovskite $CaTiO_3$ (170) and $SrTiO_3$ (270) are used, but only where their extreme temperature variation can be tolerated.

References

[1] Neugus, T., and G. J. Yeager, "Dielectric Ceramic Compositions," TransTech Patent, U.S. 5262370, 1993.

[2] Goldschimdt, V. M., "Crystal Structure and Chemical Constitution," *Trans. Faraday Soc.,* Vol. 25, 1929, pp. 253–283.

[3]   Limar, T. F., "Forming of Solid Solutions in the LaAlO3-CaTiO3 System and Their Electrical Properties," *Inorg. Chem. and Inorg. Mat.,* Vol. 5, 1969, pp. 1773–1775.

[4]   McQuarrie, M., and F. W. Behnke, "Structural and Dielectric Studies in the System (Ba, Ca) (Ti, Zr)O3," *J. Amer. Ceram. Soc.* Vol. 37, 1954, p. 539.

[5]   Cockayne, E., "First Principles Calculations of the Dielectric Properties of Perovskite-Type Materials," *J. Eur. Ceram. Soc.,* Vol. 23, 2003, pp. 2375–2380.

[6]   Davies, P. K., et al., "Communicating with Wireless Perovskites: Cation Order and Zinc Volatilization," *J. Eur. Ceram. Soc.,* Vol. 23, 2003, pp. 2461–2466.

[7]   Blasse, G. J., "Vibrational Spectra of Solid Solution Series with Ordered Perovskite Structure," *Inorg. Nucl. Chem.,* Vol. 37, 1975, pp. 1347–1351.

[8]   Koyasu, S., et al., "Dielectric Ceramic Composition for High-Frequency Use and Dielectric Material," Kyocera Patent, U.S. 5268341, 1993.

[9]   Hill, M. D., and D. B. Cruickshank, "Crystal Chemical Trends in Electrical Properties of Perovskites for Microwave Dielectrics," *Amer. Ceram. Soc.,* Vol. ACERS AMA.1A01, 2002, p. 2.

[10] Cruickshank, D. B., "1-2 GHz Dielectrics and Ferrites: Overview and Perspectives," *J. Eur. Ceram. Soc.,* Vol. 23, 2003, pp. 2721–2726.

[11] Kato, J., et al., "Dielectric Properties of Lead Alkaline-Earth Zirconate at Microwave Frequencies," *Jap. J. Appl. Phys.,* Vol. 30, No. 9b, 1991, pp. 2343–2346.

[12] Ohsato, H., et al., "Microwave Dielectric Properties and Structure of the $Ba6_{3x}Sm_{8+2x}Ti_{18}O_{54}$ Solid Solutions," *Jap. J. Appl. Phys.,* Vol. 34, 1995, pp. 187–191.

[13] Ohsato, H., et al., "Microwave Dielectric Properties of the $Ba_{6-3x}(Sm_{1-y}\backslash mbR)_{y 8+2x}Ti_{18}O_{54}$ (\mbR=Nd and La) Solid Solutions with Zero Temperature Coefficient of the Resonant Frequency," *Jap. J. Appl. Phys.,* Vol. 34, 1995, pp. 5413–5417.

[14] Shannon, R. D., "Revised Effective Ionic Radii and Systematic Studies of Interatomic Distance in Halides and Chalcogenides," *Acta. Cryst.,* Vol. A32, pp. 751–767.

[15] Ohsato, H., "Science of Tungstenbronze-Type-Like Microwave Dielectric Solid Solutions," *J. Eur. Ceram. Soc.,* Vol. 21, 2001, pp. 2703–2711.

Selected Bibliography

The best source of microwave dielectric preparation and properties is the Microwave Materials and their applications conferences:

- MMA2000, Bled, Slovenia;
- MMA2002, York, U.K.;
- MMA2004, Inuyama, Japan;
- MMA2006, Oulu, Finland;
- MMA2008, Hangzhou, China;
- MMA2010, Warsaw, Poland.

Information about abstracts of papers and subsequent published material can be obtained from the American, European, or Japanese Ceramic Societies. Papers from MMA2002 and 2006 were published in dedicated issues of the *European Ceramic Society.*

7

Metals at Microwave Frequencies

7.1 Introduction

The ability of a metal at microwave frequencies to be a low-loss conductor is a function of its bulk conductivity and skin depth. As we shall see, the bulk conductivity used to compare metals refers to their ideal state, which is completely pure, at maximum density, free of entrapped gases, and at an ideal grain size. Most methods of manufacturing or depositing metal fall significantly short of that ideal. The currents in microwave transmission conductors are concentrated near the surface, determined by the conductivity, frequency and permeability. The layer with the current is called the skin depth, and the accepted convention is to use a metal thickness of three times that thickness for any given frequency and metal to minimize losses. Skin depth is equal to $2\rho/\omega\mu_0\mu_r$ in meters, where ρ is the resistivity of the metal in ohm-meter, ω is the normalized frequency, μ_0 is the free space permeability, and μ_r is the relative permeability of the metal, normally unity for a nonmagnetic metal. Figure 7.1 shows the variation of the skin depth of copper against $\log_{10}$ frequency in hertz. At 10 GHz the value is 0.65 micron for copper and silver, compared with about 0.8 micron for gold and aluminum. Because skin depths are in the micron (10^{-4} cm) region, the conductivity is influenced by the surface finish of the metal, as this effectively increases the length of the conduction path. The chemistry

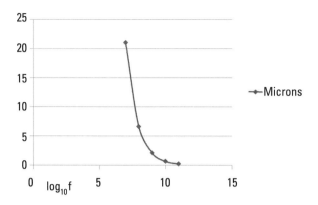

Figure 7.1   Log frequency versus skin depth thickness in microns (Y-axis) for copper.

of the surface will also be an issue if the surface is oxidized, since the oxide can absorb microwaves.

7.2   Application of Metals to Microwave Transmission Lines

Different metals used for various forms of transmission lines are shown in Table 7.1. Only copper and aluminum and their alloys are used directly in bulk. Silver, nickel, and gold are coated by various means onto other base metals. Ferrous metals prone to oxidation, with the exception of some nonmagnetic stainless steels, are always coated with more conductive metals. Rigid and semirigid coaxial outer conductors are made of copper; flexible cable outer conductors may be made of a variety of conductive woven fibers or wound wire. Inner conductors are usually copper.

7.3   Copper

Bulk oxygen-free copper is almost as low loss as silver, as waveguide, and is within 2–3% of the latter. It is available in that form over a wide range of sizes. Its principle drawback is corrosion, both directly and when in contact with other metals. As a result, it is often substituted for by copper alloys like brass or bronze, or electroplated with gold or silver, which in turn may require a nickel barrier layer to prevent alloy-

Table 7.1
Useful Metals for Microwave Transmission Lines

Element or Alloy	Waveguide	Microstrip	Triplate/ Stripline	HiQ Cavity/ Coax
Copper	x	x	x	x
Aluminium	x	—	Ground plane	x
Brass/bronze	x	—	x	Usually coated
Silver	Surface coating	x	Surface coating	Surface coating
Gold	Surface coating	x	Surface coating	Surface coating
Nickel	Surface coating	—	—	Conector shell
Stainless Steel	—	—	—	Connector shell
Invar	Must be plated	—	—	Must be plated
Kovar	—	Must be plated if ground plane	—	—

ing, potentially forming lower conductivity alloys. A copper waveguide is normally drawn or extruded, but small sections can also be electro-formed by prolonged copper plating to complex shapes using a suitable mandrel, usually stainless steel, to allow easy separation. A copper waveguide, if kept in nonoxidizing conditions, does not deteriorate and maintains its losses up to the millimetric region. Aluminum, on the other hand, forms a lossy hydrated oxide film under any environmental condition, and because of this effect and bulk conductivity differences, losses are at least 50% higher at millimetric frequencies than untarnished copper or silver. Gold has higher losses than copper but does not deteriorate in adverse conditions. Surface finish of course is also important, particularly at higher frequencies, and the data presented assumes the polished finish inside the waveguide routinely achieved by present extrusion techniques.

Copper microstrip can be realized by either thick- or thin-film techniques. Although thin film is normally the lowest loss conductor, thick-film silver and copper are close, with thick-film gold the worst. Thick-film copper requires special reducing or neutral atmospheres for processing, so thick film silver is often the best compromise between ease of processing, metal costs, and transmission line loss. The quality of alumina and its surface finish also contribute to loss. The data presented in Table 7.2 is on 0.025-inch (0.625 mm) 99% alumina with a 20-microinch (0.5-micron) surface roughness.

The lowest loss thick-film microstrip can be realized by using thick-film printing, then using thin-film etching techniques to remove the "postage stamp" serrated edge produced by the silk screen mesh, particularly above 10 GHz. This is particularly useful on thick film silver.

Thin-film copper on a microstrip is used to reduce cost compared with all-gold lines, sometimes with a protective gold layer. Copper can be vacuum evaporated or sputter coated if due care is taken to prevent oxidation in subsequent thin-film processing. To achieve reasonable thickness, thin-film copper films are copper electroplated to final thickness by either electroplating the entire surface to a suitable multiple of

Table 7.2
Composition and Conductivity of Microwave Elements and Alloys

Element/Alloy	Composition	Typical Conductivity, IACS Copper	Application
Copper	Cu	101%	WG
Low beryllium copper	99.3% Cu, 0.7% Be	Up to 52% with heat treatment, 38% average	Foil, sheet center conductor
Beta brass (> 38% Zn)	60% Cu, 40% Zn + Pb	2–28%	Machinable
Bronze (gunmetal)	Cu, Sn, Zn + Pb	10–20%	Casting
Phosphor bronze	Cu, Sn, P	~25%	Foil, sheet
Alpha brass (< 38% Zn)	70% Cu, 30% Zn; or 95% Cu, 5% Zn; or 98% Cu, 2% Zn	27%; or 44%; or > 75%	Stamping, drawing, extrusion
Aluminum	Al	57–61%	WG
Al/Mg, or Al/Mg 40E	99% Al, 1% Mg, or 93.9% Al, 0.6%, Mg, 5.5% Zn	50%, or 35%	Machinable investment casting
Aluminum silicon	Up to 12% Si	26–39%	Investment casting
A380	Al, Si,Cu, Zn	26– 35%	Pressure casting

the skin depth at the microwave frequency, then etching using conventional photo resist lithography, or alternatively using photo resist developed to expose the required circuit, then plating up through the resist as a boundary followed by resist and thin-film removal to create the required circuit. The latter is often referred to as pattern plating. Another source of variation in loss is the effect of copper metal surface roughness on the conductor contribution to insertion loss on copper laminate, where of course the loss being measured is determined by the surface finish of the copper foil or plating on the dielectric to form the microstrip conductor [1, 2]. Figure 7.2 shows the loss in dB/meter for three different copper surface finishes, up to 12.5 GHz, with the dielectric losses discounted, for a copper clad 50-ohm microstrip. The dielectric thickness used was 77.5 microns. Laminate manufacturers typically have to trade copper losses from surface finish against adhesion to the laminate surface [3].

Copper foil is used for center conductors in triplate stripline devices. To maintain rigidity, beryllium copper or other copper alloys like brass or phosphor bronze are used. Because some copper alloys have only 10% to 25% of the conductivity of pure copper, these alloys may require electroplating with copper or silver for high-Q cavities or low insertion loss triplate center conductors. Conductivity is largely a function of the major additive. Low beryllium copper, at 0.7% Be, is much more conductive than alloys with 2% Be; low Zn alpha brasses are

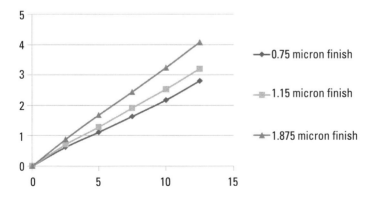

Figure 7.2 Conductor loss (dB/m) contribution for three surface finishes of copper laminate, versus frequency, GHz (X-axis). (*After:* [3].)

much more conductive than common beta 40/60 brasses (Table 7.2). The metals in the table are compared with the internationally accepted value for copper (IACS), assumed to be 100%.

Electroplating with copper has issues because of stress induced in the copper film by, for example, additives to the plating solution. A further issue is grain size; annealing of copper plating in an inert atmosphere may be necessary to achieve full conductivity and relieve stress.

Selective plating of copper is possible to minimize cost. It is normally done by suitable selective masking prior to plating. Electroformed copper has the same microstructural issues but is extremely useful for small scale, complex, or dimensionally very precise shapes.

Soldering or wire bonding to copper can be an issue because of oxidation. More aggressive cleaning and solder fluxing are required, and copper may require gold plating to facilitate gold wire bonding. Gold plating on copper normally requires a nickel barrier layer to prevent alloy formation and subsequent poor adhesion or conductivity.

Copper is not normally cast for microwave purposes, but copper bronzes are; cast waveguide component bodies are one example.

7.4 Aluminum

Extruded aluminum waveguide is used for weight reduction. It is often given protection against moderate environmental conditions with phosphate-chromate conversion coatings, which slightly increase loss and surface resistivity, but these coatings easily break down with pressure or mild abrasion if good grounding is required. Anodizing is not normally an option for conducting surfaces because of the hydrated oxide films that form. Hard anodizing, which is carried out at low temperatures, is useful for creating robust insulating surfaces, but cannot be used for microwave conductors as the hydrated oxide film is even thicker. Selective anodizing may well be an option if required for some structures.

Aluminum can be sputtered on to metallic or ceramic surfaces to give a light, highly conductive layer, including on to ferrite for filled waveguide phase shifters and on dielectric for fully enclosed resonators. Aluminum cannot be electroplated, and sputtering or evaporation may be the only option for some applications.

Casting of aluminum is a very common operation for both waveguide and cavity structures, and housings and bodies for triplate and microstrip circuits. Casting may be a gravity die, "lost wax," or pressure die operation, each with its own alloys and techniques. The conductivity of aluminum castings is largely determined by the silicon content, up to a maximum of 12% Si (Table 7.2). Essentially, loss of conductivity must be traded against ease of casting with higher Si content. High Si aluminum is more difficult to electroplate, and special surface preparation is required.

Deep drawn pressing of aluminum "supermetals" has changed the range of application of thin-walled enclosures, and these can be an option compared with machined or cast aluminum.

7.5 Silver

Because of its cost, silver is rarely used as a bulk metal in microwave applications. Most silver is applied by electroplating on to metals. Special techniques have been developed to plate onto aluminum, involving removing the oxide skin and electroless deposition before standard plating. Cast aluminum contains high levels of silicon, which make this initial coating more complex. Silver is rapidly tarnished by atmospheric sulfur, particularly, and although this process can be inhibited by surface treatments it limits the duration of optimum loss in high-Q cavity filters. Waveguide components electroplated with silver can be rhodium "flash" plated to improve corrosion and wear resistance. It is possible to selectively plate silver on high-Q areas of aluminum bodies to minimize the additional cost, by appropriate masking techniques prior to plating.

Thick-film silver is used on ceramic dielectrics and magnetic materials. Because of lower conductivity, palladium silver is not generally used at microwave frequencies. In general, fritless silver pastes are used, as glass also reduces conductivity. As a result, bond strengths can be weak and soldering less effective, so pure silver fritless conductors are used where soldered or brazed joints are not severely stressed. Soldering or brazing to thick-film silver is normally carried out with silver containing solders or brazes to prevent severe silver diffusion from the thick-film weakening the joint.

As mentioned previously, thick-film silver microstrip lines will show a serrated edge due to the silk screen mesh (which is usually made of fine mesh stainless steel). This can be removed by photolithography to improve losses due to the edge currents. Silver microstrip circuits can be subject to "whisker" (dendritic) growth in the presence of moisture and a DC voltage, for example for voltage biasing of semiconductor devices. This can be avoided in some cases by using thick-film insulating dielectric layers over the silver lines, although care is required to avoid significant additional loss from the dielectric of the insulator.

7.6 Gold

Gold is almost exclusively used in thin-film or electroplated form because of its cost. Its main advantage over silver and copper is corrosion resistance because of its chemical inertness and the ease with which gold wire or tape can be thermocompression or ultrasonic bonded to it. Gold and palladium gold thick-film inks are used occasionally for these reasons for microstrip lines.

Thin-film pattern plated gold can behave in different ways. Some gold-plating solutions will give a mushroom type of shape to the microstrip line cross-section, while others will give a columnar structure with a squarer cross-section, which is usually better. Gold plating on copper and brass requires a barrier layer to prevent alloy formation through diffusion, nickel being the usual choice.

7.7 Relative Losses of Metals in Microstrip and Waveguide Transmission Lines

Table 7.3 shows the loss on 0.635 mm (0.025 inch) alumina, finished to 0.4 to 0.8 micron (10–20 microinches), for a 50-ohm line at 2 GHz, where line width is approximately also 0.635 mm. Note that the losses are in all cases much higher than predicted from theoretical bulk metal conductivity.

Table 7.4 shows the relative theoretical losses of bulk copper and aluminum waveguide (WR90 at 10 GHz and WG28 at 35 GHz) and the effect of plating with silver and gold. Although the theoretical losses are already higher at 35 GHz than 10 GHz, in fact the difference

Table 7.3
Losses of Different Metals on 50-Ohm Microstrip Lines on Alumina

Metal	Thick-Film Loss	Thin-Film Loss	Theoretical Loss
Silver	0.15 dB/cm	—	0.04 dB/cm
Copper	0.16 dB/cm	—	0.04 dB/cm
Gold	0.23 dB/cm	0.12 dB/cm	0.06 dB/cm

Table 7.4
Theoretical Losses of Different Metals in Waveguide at Different Frequencies

Metal or Plated Metal	Loss in dB/m at 10 GHz, WR90	Loss in dB/m at 35 GHz, WR28
Copper	0.11	0.54
Gold plating	0.13	0.64
Aluminum	0.14	0.69
Silver plating	0.105	0.52

can be much higher in actual waveguide because of surface roughness, and in the case of aluminum, the oxide film, and as a result actual losses can be 10 times higher or more. This can be minimized by plating the aluminum with gold or silver.

7.8 Nickel

Nickel electroplating can be used in some applications where losses are not critical and due account is taken of potential intermodulation effects or corrosion when in contact with other metals. Although the permeability of magnetic metals affects the skin depth of metals like nickel, its conductivity allows it to be useful for connector housings, for example. Nickel plating can be stressed internally, and is normally used with a copper undercoat when plated to other metals to improve adhesion. A thin chromium layer is often added to reduce the slight tarnishing of pure nickel plating.

7.9 Steels

Nonmagnetic stainless steel, unplated, is used where losses are not critical, for example for connector housings. It has relatively poor conductivity, which limits its usefulness. Plating onto stainless steels requires similar techniques to Invar (Section 7.11).

Mild steel is often used for magnetic devices like circulators and isolators to improve the effectiveness of magnets used to bias the device, by completing the magnetic circuit around the device. It has poor conductivity and requires plating if it also forms part of the microwave circuit, but it is cheap and easy to machine.

Special steels and other ferrous alloys are used as temperature compensating elements in magnetic circuits; others as low thermal expansion alloys for temperature-stable resonant cavities; and yet others for matching thermal expansion with alumina, glass, and magnetic or dielectric ceramics where the alloy and the other material are joined by soldering, brazing and gluing.

7.10 Magnetic Temperature-Compensating Alloys

These typically come in a range of Curie temperatures [4], and can be used in shunt or series in magnetic circuits to match the field of the biasing magnet with the changing internal magnetization of the ferrite element over temperature. Although nominally of the same composition, there are several types with different permeabilities and permeability against temperature slopes, brought about by different thermal treatments [3].

7.11 Metal Alloys with Low or Zero Expansion Coefficient

These come in a series of nickel-iron alloys, with Invar, also known as Nilo 36, having essentially zero expansion near room temperature. Invar has been used for temperature-stable waveguide cavities and filters and many other cavity applications. Waveguide filters made of Invar are typically plated with copper or silver at lower microwave frequencies where the waveguide is large, or gold at millimetric frequencies. Invar post filters use invar or steel rods brazed or, less often, soldered into the waveguide prior to plating. Special cleaning methods like water or air sandblasting of these assemblies prior to plating are necessary. Tuning screws are typically made of mild steel, which, like all tuning screws made of ferrous alloys, must be plated with the same metal as the filter is plated with to avoid intermodulation due to dissimilar metals and low Qs due to the poor conductivity and magnetism of steel and Invar. Note that because of the high surface area and potential for thin or worn plating on screws, special attention must be given to electroplating of tuning screws. Some plating solutions may not "throw" into the depth of the screw thread, potentially creating plating thickness below the required number of skin depths in the plated metal.

Kovar is used to match the expansion of alumina or similar ceramics, and is typically used as a carrier for microwave integrated circuit substrates of these materials. If it forms part of the ground plane it is usually plated, or it may be plated to allow soldering or brazing to the ceramic.

The properties of this group of metals are shown in Table 7.5.

A summary of bulk-forming methods and surface deposition methods for metals commonly used at microwave frequencies is given in Table 7.6.

7.12 Metal Plating on Plastics

Two of the issues with metal microwave parts are firstly weight, and secondly, in the case of metals other than invar, lack of temperature stability mechanically, resulting in frequency drift of frequency-selective

Table 7.5
Thermal Properties of Ferrous Alloys

Alloy	Composition	Expansion Coefficient	Curie Temperature	Source
Nilo K (Kovar)	Ni 17:Co 24:Fe 54, 5% other	3–5 ppm/°C	435°C	Carpenter
30 type 2,3,4	Ni 30:Fe 69, 1% other	6 ppm/°C	71°C	Carpenter
Invar (Nilo 36)	Ni 36;Fe 63, 1% other	1 ppm/°C	180°C	Carpenter
Mild Steel	Fe: Carbon	12 ppm/°C	768°C	Various

Table 7.6
Bulk-Forming Methods and Surface Deposition Methods for Various Metals

Metal/Alloy	Wrought	Cast	Plating	Thick Film	Thin-Film Evaporation	Sputter
Copper	x	Rare	x	x	x	x
Aluminium	x	Rare	—	—	x	x
Silver	—	—	x	x	x	x
Nickel	—	—	x	—	x	x
Gold	—	—	x	x	x	x
Copper alloys	x	x	Rare	—	x	x
Aluminium alloys	x	x	—	—	x	x

waveguide or cavity devices when temperature cycled. Weight is an issue for satellite payloads, but more recently the use of tower-top near-antenna filters in cellular infrastructure base stations, tower-mounted waveguide-based point-to-point radios, and antenna feed structures has brought weight reduction into consideration for these applications.

The early use of metalized plastic took the form of winding un-cured carbon fiber/epoxy tape on to a mandrel, curing the tape with a high-pressure "prepreg" process to remove voids, removing the mandrel, then using electroless nickel plating to form the initial plastic/metal interface, followed by electroplating to the desired thickness. Prepreg is the lay-up process used by reinforced laminate manufacturers. This, however, did not result in sufficiently temperature-stable parts. Later, a different process used construction of the desired shape out of glass fiber reinforced high-temperature plastics like Ultem (Chapter 4) by gluing fabricated parts together without a mandrel, then coating with electroless copper followed by electroplated copper. Again temperature stability was poor, but weight reduction of 30% to 50% was claimed over aluminum parts of the same internal size. This process was followed by processes using mandrels in unique ways. One used a mandrel electroplated to the desired internal finish, which was then wrapped with carbon or glass fiber polymer composite. The whole assembly was then cured, and the mandrel removed, leaving its plating attached to the polymer composite. ESTEC (part of the European Space Agency) jointly developed a process with Polymer Kompositer AB using carbon fiber reinforced tape on a mandrel cured using a high-pressure prepreg process, followed by electroless and electroplating after mandrel removal [5]. This process was said to be more stable (expansion was only 1 ppm/°C) than Invar (2 ppm/°C) after prolonged temperature cycling of the same device design. By contrast, the same device in aluminum was 23 ppm/°C. Polymer Kompositer is currently using a fusible tin alloy mandrel, which is melted out after being coated with a reinforced plastic. The remaining hollow shell can then be plated [6].

Processes like these are increasingly being used on antenna feeds, filters, and waveguide sections where lightweight and reasonable temperature stability is required.

References

[1] Meinel, H., et al., "Optimization of Thin and Thick Film Technology for Hybrid Microwave Circuits," *Electrocomponent Sci. Tech.* Vol. 6, 1977, pp. 143–146.

[2] Johnson, R. W., et al., "Advances in Thick Film Conductors for Microwave Integrated Circuits," *Electrocomponent Sci. Tech.* Vol. 4, 1980, pp. 214–216.

[3] Carpenter Technology Corporation, www.cartech.com.

[4] www.arlon-med.com.

[5] Crone, G. A. E., et al., "High Performance Metalized Composite Antenna Feeds and Microwave Components," ESA-ESRIN ID/D, ESA Internal Report, 1995.

[6] Polymer Kompositor, www.polykomp.com.

Selected Bibliography

Moreno, T., *Microwave Transmission Design Data,* Norwood, MA: Artech House, 1989.

Ross, R. B., *Metallic Metals*, London, U.K.: Chapman Hall, 1968.

Wadell, B. C., *Transmission Line Design Handbook*, Norwood, MA: Artech House, 1991.

8

Ferrite Devices

8.1 Introduction

Microwave ferrite devices cover three basic types, depending on the degree of DC magnetic bias applied. These are those operating below ferromagnetic resonance, those at resonance, and those above resonance. Figure 8.1 shows the absorption behavior of a ferrite sphere at 9.3 GHz. Low field absorption loss can be seen at the zero field point and is dependent on the ferrite magnetization. For magnetizations below about 3,300 gauss, and using the demagnetization factor for a sphere (1/3) there is a plateau of low absorption between low field loss and ferrimagnetic resonance, which is used for below-resonance devices. In this case, the resonance is at 3,000 gauss of applied field. Well above resonance, losses fall again to low levels, and in this case device operation above resonance is possible; however, at 9.3 GHz, this requires a strong field. For higher magnetizations at that frequency, the low field loss and resonance absorption loss begin to merge, until there is no below-resonance bias point where there are low losses. This absorption behavior scales over all microwave frequencies, with important consequences for both material selection and mode of operation.

Devices also fall into two major categories: those using a ferrite junction of three or more ports, and those based on two-port linear devices.

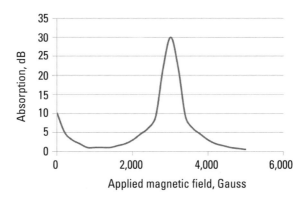

Figure 8.1   Magnetic bias (X-axis), as an applied DC magnetic field, versus absorption, showing ferrimagnetic resonance.

Most ferrite devices above or below resonance are nonreciprocal. This comes about because, when a ferrite at RF frequencies is biased magnetically, the RF is polarized in two different opposing senses, each with its own characteristic mode, which, in the case of a junction circulator, is two TM modes of opposite polarization. By arranging the input and output ports of a device, one mode may appear at a different output port from the other, creating possible nonreciprocal operation. This is typically done in a junction device to create an n-port circulator, where n is typically 3 but can be 4 or more. A three-port circulator is shown in Figure 8.2, illustrating the principle. Port 3 is isolated when the modes there differ in phase by an odd multiple of π.

The analysis of a three-port junction was initially carried out by Bosma [1], followed by Comstock and Fay [2], Davies and Cohen [3], and later Helszajn, the latter comprehensively describing n-port junctions in a series of articles and books (see the Selected Bibliography). Most of the evolution of this analysis is covered in Linkhart [4], and the purpose of this chapter is to cover the factors involved in choosing materials rather than the junction analysis, although the two are obviously related. Circuit analysis gives solutions covering the frequency and bandwidth of devices, and their dimensions, but the loss mechanisms require certain assumptions about the magnetic absorption characteristics, which are not valid in many instances. The analysis of the behavior of the material from the physical rather than the circuit aspect was largely carried out by Schloemann [5] to take into account

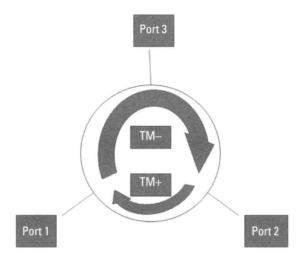

Figure 8.2 Operation of a three-port circulator.

magnetic saturation, resonance, and nonlinear behavior in ferrite. Numerical analysis of inhomogeneously magnetized ferrite junctions was carried out by Krowne [6] in an attempt to reconcile accepted circuit theory with the behavior of ferrite in junctions. In this chapter, aspects from circuit and physical aspects of ferrite in a junction will be compared with actual experience of a wide range of materials in devices.

8.2 Below-Resonance Junction Devices—Selecting the Correct Magnetization

In device design, one of the first considerations is bandwidth and hence magnetization. Low field loss as it is known is caused by a lack of saturation at a given frequency, and the frequency threshold (Ft) below which low field loss is usually said to occur is when $Ft < N \times \gamma.4\pi Ms \times C$. While different values can be proposed for demagnetization factors N and constants C, the equation assumes that the ferrite is uniformly saturated, an essentially impossible condition in a real device unless the ferrite is spherical or ellipsoidal, as these are the only shapes that produce a uniform internal DC field.

Circuit theory developed from Fay and Comstock [2] essentially suggests that the fractional bandwidth $\omega^+ - \omega^-/\omega$ of junction devices at center frequency ω is determined by the phase angle of each of the two counter-rotating TM modes, of frequency ω^+ and ω^- and of opposite polarization, being 30°, relative to the zero bias condition at the center frequency, such that $\omega^+ - \omega^-/\omega$ is proportional to 1/loaded Q. The mode splitting set up by the biasing field can also be related to the ferrite tensor permeability factors, κ and μ, such that 1/loaded Q is proportional to κ/μ. The loaded Q is therefore a function of bandwidth and magnetic loss, if dielectric losses are neglected. Below resonance, there are both low field losses and resonance losses to consider. Since the saturation magnetization of a ferrite changes with temperature, it is also apparent that the internal field in the circulator must also be adjusted to keep the same bandwidth or splitting of the modes. This involves actually reducing the internal field as the magnetization increases, potentially leading to an unsaturated ferrite and hence low field loss if initially close to that point at room temperature. Conversely, if the magnetization decreases, the magnetic field must be increased, potentially bringing the ferrite closer to ferromagnetic resonance in below-resonance operation.

The repercussions are considerable when we consider the sum of these factors. The choice of magnetization of the ferrite must take account of bandwidth, temperature range, and the fact that the ferrite is nonuniformly magnetized because of its shape. In addition, as we shall see, the lack of uniformity inside the ferrite may be the main source of nonlinearity if as a result the ferrite is not saturated everywhere, or is too close to resonance under certain operating conditions.

It may seem logical then to keep the magnetism as low as possible for a given frequency. However the corresponding lack of splitting may either result in insufficient bandwidth or bring the ferrite too close to ferromagnetic resonance, resulting in higher losses. We can get some idea of the limits of what is possible by looking at practical octave bandwidth circulators with matching transformers. Figure 8.3 shows the optimum $4\pi Ms$ at room temperature for below-resonance octave devices, where insertion losses at the band edges are less than 1 dB, and isolation and return loss are about 20 dB. Insertion losses are higher at the low band edge because unless the optimum field is chosen, that is where higher losses from low field loss and resonance will appear. If

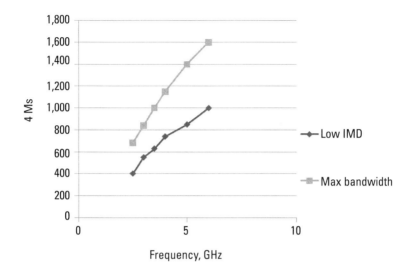

Figure 8.3 Difference between $4\pi Ms$ selection (gauss, Y-axis) for bandwidth, and selection for low IMD and optimum power handling, against center frequency.

we then contrast this with the lowest loss device possible at the same center frequency, much lower $4\pi Ms$ vaules are possible, but with less bandwidth. However, nonlinearity in terms of power handling and intermodulation (IMD) are much better, simply because although all of the ferrite is still not uniformly saturated because of its shape, it is much more saturated than its higher $4\pi Ms$ counterpart.

Work on shape-related saturation and its effect on bandwidth, power handling, insertion loss, and IMD is scattered throughout the literature. Gross demagnetization factors for shapes have been around for many years, but Pardavi-Horvath [7, 8] has shown the distribution of the demagnetization tensor across a ferrite disk (Figure 8.4) and rectangular slabs of finite thickness, illustrating the extreme effects at the edge of disk and at the edges and corners of the slab.

Schloemann [9] was able to extend the loss bandwidth of circulators using hemispherical ferrites to shape the internal DC field, and How [10] suggested the same demagnetization factors reduced the IMD of circulators. Later, Muira [11] was able to show that an ellipsoidal ferrite resonator had significantly better IMD than a disc of the same material under the same power and frequency conditions.

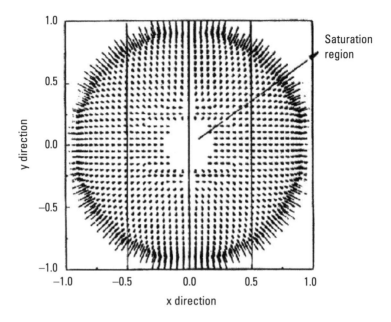

Saturation
region

y direction

x direction

Figure 8.4   Demagnetization vectors in a perpendicularly biased ferrite disk.

Another approach is to use concentric rings of ferrite, with high magnetization at the center of the junction and lower magnetization at the edge of the composite disk, the object being to saturate each ring. Krowne [6] has analyzed and built models of such devices, and Shloemann [12] later refined this approach, giving more generalized design data.

Concentric rings of ferrite may be the most practical way to realize optimum insertion loss bandwidth and low nonlinear behavior in junction devices. The broad conclusion that can be drawn is that insertion loss, power handling, and IMD can be improved by better saturation of the whole of a ferrite disk in a below-resonance junction device, making the choice of magnetization a critical factor in device design. These factors must be traded off against bandwidth, which typically requires a higher $4\pi Ms$.

8.4 Magnetization Against Temperature

Attempts by Weiss to predict the magnetization against temperature behavior of magnetic materials were not successful in predicting the complex behavior of multisublattice systems of the type found in ferrimagnetic materials, and currently actual measurements of magnetization versus temperature are the only useful source of data (Chapter 12). With the exception of systems with compensation points like Gd-substituted garnets, all other ferrimagnetic systems show a monotonic fall in magnetization from absolute zero to the Curie temperature. From a device characterization point of view, a below-resonance circulator is most temperature stable with an invariant $4\pi Ms$ and magnetic field, the latter being the only realistic possibility as magnets exist that do not change significantly, at least relative to the ferrite in the junction. We can break down the slope of the magnetization to that for below room temperature and that for above, for the commonest ferrites used in the full frequency range of below-resonance devices (see Table 8.1).

Table 8.1 requires some qualification. The temperature slope chosen is thought to be the best compromise between insertion loss, particularly at low temperature, and temperature stability. For devices where the temperature range is wide, but the insertion loss is less important, higher Gd (see Chapter 1) containing garnets of the same magnetization at room temperature are available giving better stability. Similarly, Ni ferrites have higher Curie temperatures than the corresponding MgMnZn ferrites of the same magnetization, but Ni ferrites close in magnetization to pure Ni ferrite may have higher losses (Chapter 2). NiZn ferrites do not have this problem. To minimize insertion loss, spinel ferrites containing Co and garnets containing Ho and other relaxers should not be used in low-power devices. In general, Li ferrites are not used much in nonswitching devices, so these are not included in this section.

A further qualification is the choice of magnetization for a given frequency. As explained earlier, the lowest frequency is a function of $\gamma \times 4\pi Ms$ in megahertz. However, this expression must be factored for demagnetization factor and power, insertion loss, and IMD. If these are important, the design should be factored toward the highest frequency of the range. There is really no maximum frequency for any $4\pi Ms$, the

Table 8.1
Typical Below-Resonance Ferrites Used in Low Power Junction Devices at Moderate Bandwidth[*]

$4\pi Ms$ Gauss	Composition	$4\pi Ms/T$ %/ °C, −60°C to RT	$4\pi Ms/T$ %/ °C, RT to +85°C	Curie Temperature (°C)	Frequency Range (GHz)*	Source TT = Trans Tech
300	Al-YIG	−0.4	−0.5	120	1–2	TT,Temex
400	Al-YIG	−0.4	−0.5	130	1.3–2.5	TT,Temex
550/600	GdAl-YIG	−0.1	−0.25	180	1.6–3	TT,Temex
680	GdAl-YIG	−0.1	−0.2	200	2–4	TT,Temex
800	GdAl-YIG	−0.05	−0.15	240	2.6–5	TT,Temex
1,200	GdYIG	+0.05	−0.1	280	3.2–6	TT,Temex
1,600	GdYIG	−0.2	−0.2	280	5–10	TT,Temex
1,780	YIG	−0.2	−0.2	280	5.5–11	TT,Temex
2,000	MgMnAl	−0.22	−0.22	290	6–12	TT,Temex
2,500	MgMn	−0.22	−0.22	320	8–16	TT,Temex
3,000	MgMnZn	−0.27	−0.27	240	9–18	TT,Temex
4,000	NiZn	−0.2	−0.2	470	12–24	TT,Temex
5,000	NiZn	−0.21	−0.21	375	15 upwards	TT,Temex

*Assumes 20% bandwidth.

recommendation in the table marked [*] is based on achieving at least 20% bandwidth. For many applications, including point-to-point radio waveguide isolators, the required bandwidth is relatively low and hence lower $4\pi Ms$ materials can be used successfully.

Note that for above resonance, a completely different set of choices must be made.

8.5 Insertion Loss Considerations Below Resonance

The prime ferrite contributions are to dielectric and magnetic loss. In stripline devices below resonance, dielectric losses should ideally be less than 0.0002 at 10 GHz, and most commercial garnets achieve this level. Typically those containing only trivalent ions [Y, Gd, Fe, Al, In and most rare earths (see Chapter 1)] are less than 0.0001. Those containing ions with nontrivalent ions (Ca, V, Zr, Ge, Sn) are more variable and require very tight stoichiometry control to achieve the lower value. Values above 0.00025 or so will affect the insertion loss. Typically spinel ferrites are not used in below-resonance stripline junction devices where insertion loss is critical and a garnet of the same magnetization is available. Spinels are more usually used in stripline broadband devices at higher frequencies for radar and instrumentation, where their losses, which are in the 0.0002 to 0.0005 range, are more tolerable. They are, however, used extensively in waveguide junction devices, where at frequencies in the 3- to 20-GHz range there are usually relatively few other contributors to insertion loss other than the ferrite, although the waveguide itself becomes an issue above those frequencies (Chapter 7).

Magnetic loss is much more complex. Assuming we have chosen the correct magnetization, low field losses should not be an issue. At high microwave frequencies, at the low end of the frequency band of interest, there should be enough DC field separation between both low field losses at low fields and the tail of the ferromagnetic resonance at high fields. This gives a wide choice of garnets or spinels with relatively wide resonance line widths, which can be several hundred Oersted for spinels. It should be noted that the highest frequency that low field loss can occur at is actually $\gamma \times 4\pi Ms + K1/Ms$, so the line width is a consideration since it is essentially $K1/Ms$ plus microstructural contributions (Chapter 1). However, we need to differentiate between spinels and garnets where the magnetocrystalline anisotropy comes from distortion of the Fe^{+3} ions in their crystallographic environment or from porosity and low magnetic loss second phase material, and those which contain fast relaxers (Chapter 1) or moderate relaxers like Gd, where the intrinsic magnetic loss may be high because significant amounts of Gd are used for temperature compensation. Those with relaxers will still have losses proportional to their doping level away from both low field loss and resonance and represent an insertion loss limitation independent of bias.

Below-resonance stripline devices have dielectric loss contributions from glue (Chapter 4) used to position the ferrite or tuning dielectric, the tuning dielectric itself (Chapters 4, 5, and 6) and the dielectric and glue present in a ferrite/dielectric composite assembly if glue is used. Silicones are generally used where a significant amount of glue is needed (some silicone glues are used as the tuning dielectric), but these are not lossless. Note that glues should take account of expansion differences between the materials employed. Again, flexibility requirements, particularly at low temperature, favor silicones over more rigid alternatives.

The body of the device, when it is used as a ground plane, and the center conductor are typically silver plated (Chapter 7) to minimize losses. Values approaching 0.1 dB per junction are possible for a well impedance-matched device if due precautions are taken. Similar values are possible in waveguide (WG) junctions in the 3- to 20-GHz range with suitable choice of ferrite, dielectric, metals, and glues.

8.6 Power Handling in Below-Resonance Junction Devices

Junction devices suffer from the disadvantage of relatively compact structures, from a power handling point of view. Because below-resonance devices readily become nonlinear in power absorption as peak power rises (Chapter 1), it may be necessary to consider using fast relaxers in devices in the 10W (40 dBm) and upwards range. This threshold is extremely variable and difficult to predict, because it depends on the magnetization, which should be as low as possible, the DC bias, the possibility of subsidiary resonance, the RF field intensity due to the ferrite thickness, and the fact that heating itself through linear power absorption can bring the threshold for nonlinearity to a lower peak-power level, because the spin wave line width δ Hk, drops with increasing temperature. This chain of events manifests itself as thermal runaway, with catastrophic failure a possibility if not controlled. It follows that very conservative margins should be applied to the onset of nonlinearity. In addition, the microstructure of the ferrite itself adds variability. In this situation, porosity and nonmagnetic second-phase materials actually help delay the onset of nonlinearity, and small grains are preferable to large ones.

In the author's experience, the best way to follow such a conservative strategy is to choose a suitably low $4\pi Ms$ material, without fast relaxers, then measure its peak-power handling at low mean power (by using a low pulse repetition rate source if available) to establish the critical threshold point, Pcrit., at the necessary DC bias for meeting all other relevant parts of the specification in the junction. Using the published values of δ Hk, it is then possible to estimate the peak-power handling of undoped and fast relaxer-doped materials, using the equation determining the critical RF field required for the onset of instability,

$$h\,crit. = \omega\delta\,Hk/\gamma \times 4\pi Ms \qquad (8.1)$$

where Pcrit. = (hcrit.)^2. This estimate should then be verified by measuring the peak power of the chosen doped material at low mean power and at the actual mean power to be used. If the peak power is still not met, this is repeated at higher doping levels. The strategy then is to attempt to improve the mean power handling by heat sinking the ferrite and its surrounding area as much as possible. The end point is testing the device at the highest peak and mean power at the highest temperature of use to ensure the specification is met with at least a 20% margin. Heat sinking can be helped by using thermally conductive metals and alloys, which have a high specific heat (usually proportional to density) and the use of thermally conducting dielectric if required (Chapters 6 and 7). External loads may be required for isolators, using black body radiation and high surface area principles in the form of painted fins and similar techniques. Silicone-based greases may be necessary to ensure good thermal contact between surfaces. In extreme cases, water cooled loads may be required.

8.7 Intermodulation in Below-Resonance Junction Devices

The choice of $4\pi Ms$ is similar to that found with peak power, which is that of picking a value as low as possible to meet the IMD specification. The peak power requirement should be met first, if possible by using the lowest possible magnetization without doping. Improving the IMD is not possible using fast relaxers as the problem is normally lack of saturation in the ferrite (Section 8.2), not power nonlinearity if this has already been met by the choice of magnetization. It may be possible to increase the DC bias without bringing the tail of resonance in at low frequency, which will help IMD. Most issues, however, relate to the complete DC bias circuit, which may contain magnetic paths designed to minimize magnet size. The designer needs to consider the proximity of the return path (as in Figure 8.5, where the DC return field is in the opposite direction to the magnet polarity), the size and shape of the magnet pole piece, and the presence of any other magnetic element in the device, including compensating steels and the body itself if a magnetic alloy like steel is used (Chapter 7). One method that is effective is the use of a high dielectric constant ring around the ferrite (Figure 8.5), which keeps the center frequency constant by substituting dielectric for ferrite, but reduces the proximity of return path magnetic metal while making the DC field uniformity easier to achieve using pole pieces, which can then more than cover the ferrite. If we consider the position of the magnetic field dipole at the edge of the disk due to its demagnetization factor [7, 8], the need to saturate the ferrite at its edges, and the radial nonuniformity of the DC field created by the magnet/pole piece and return path circuit, we have a "perfect storm" situation from the IMD point of view. In addition, the influence of the RF field at the disk edge should be considered [13].

The use of concentric rings of ferrite of reducing magnetization from the center outwards [6, 12] can clearly positively contribute to improving this situation (Figure 8.6), as expensive shapes such as spheres or ellipsoids of ferrite to improve internal DC field uniformity are not usually an option and are difficult to impedance match to in any case. In some cases, thickening the ferrite may help somewhat by reducing RF field intensity. Attempts to improve IMD by adjusting the center conductor geometry have been reported [13].

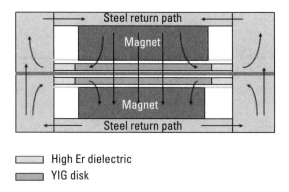

Figure 8.5   Magnetic return paths and concentric dielectric rings in a junction.

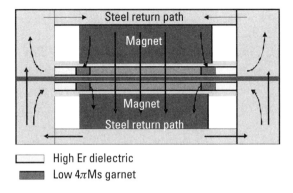

☐ High Er dielectric

▨ Low 4πMs garnet

Figure 8.6 The use of concentric rings of dielectric and ferrite of low 4πMs.

8.8 Microstrip Below-Resonance Devices

Microstrip-based below-resonance devices on ferrite substrates present a further series of issues. These relate to the increased field intensity around a microstrip, and the difficulty of uniformly saturating a square or rectangular substrate containing ferrite junction(s) asymmetrically placed on its surface. As a result, such devices are not suitable for low IMD requirements, and have very low peak-power handling, sometimes in the 10-mW region. Power handling can be improved by suitable relaxer doping. Insertion loss is much higher than stripline or WG junction devices, usually because of a combination of these factors.

Switching below-resonance WG junction devices are possible using the BH loop properties of ferrite. Material selection for these is based on the above principles plus the BH properties explored in Section 8.9.

8.9 Below-Resonance Linear Devices

Although this class of device is fundamentally the result of the behavior of a rectangular ferrite slab in a waveguide or a stripline, or sometimes a microstrip or coplanar device, it is possible to make nonreciprocal devices by suitable additions of couplers or "magic tees," as discussed by Linkhart [4]. Material selection has been studied most in WG devices in the form of differential phase-shift circulators or isolators [14]. A special case is the use of switching or latching ferrites to create magnetically adjustable phase shifters or ferrite switches.

The operation of the phase shifter is determined by the difference in phase of the oppositely polarized circularly polarized waves in the ferrite slab in WG, such that any section achieves at least 90 or 180 degrees of phase shift, depending on the design. The differential phase shift per unit length of ferrite is then a function of the slab dimensions and position in the waveguide and the familiar term κ/μ (Section 8.2). This means the selection criteria for a differential phase shift circulator ferrite is similar to a junction device, but with an important additional degree of freedom. By choosing a very low 4πMs, thin slabs, and suitable bias field, it is possible to make the device as long as the available space permits, thus improving the peak-power handling (low 4πMs) and also mean power handling by heat sinking the ferrite on to the WG wall because the area of ferrite on waveguide has increased. This means that a wider but lower range of 4πMs available is greater than a junction device, where the geometry is constrained by the TM mode. WG differential circulators and isolators have been built with power handling comparable with the WG itself. The design criterion for choosing the correct ferrite is the same, that is selecting the 4πMs to give the necessary phase shift per unit length, testing an undoped ferrite of known spin wave line width to measure the nonlinear point, calculating the spinwave line width required from a doped material, then retesting at both peak and mean power at the highest operating temperature. Doped garnets are available in the range 550 to 1,780 gauss, usually with Ho^{+3} but Co^{+2} can be used (Chapter 1). Magnesium ferrite can be doped with Co^{+2}. Nickel ferrites have some intrinsic relaxation from A-site Ni^{+2}, which is reduced when Al or Zn are used to decrease or increase the magnetization, and so Co^{+2} is required for most Ni ferrites if greater power handling is required.

8.10   Switching and Latching Devices

The material requirements for devices where continuously or digitally variable phase shifting is needed are similar to differential phase shifters, with some additional features. These devices usually operate using a rectangular cross-section toroid in a waveguide. Instead of a magnet to provide DC magnetic bias, a wire passing through the toroid can magnetize the ferrite using a current pulse of sufficient amplitude. The ferrite toroid retains this magnetization, because it is a complete magnetic circuit, as the remanence, Br, determined by the B/H loop (Figure 8.7).

Since the remanent field in the toroid determines the phase shift along its length, the phase change in the toroid will be proportional to the amplitude of the pulse. The phase shifter can be either a continuous length of ferrite that is varied in phase shift (analog) or a series of shorter sections of predetermined phase shift "bits," which are switched in and out (digital).

This introduces two important parameters, the Br/Bm ratio from Figure 8.7, which determines how much remanence is retained for a given current pulse, and magnetostriction set up by stresses on the toroid either by machining it to size, or by external stresses set up by the method of fixing the toroid to the waveguide. Magnetostriction can reduce or increase the Br/Bm ratio, depending on the material and

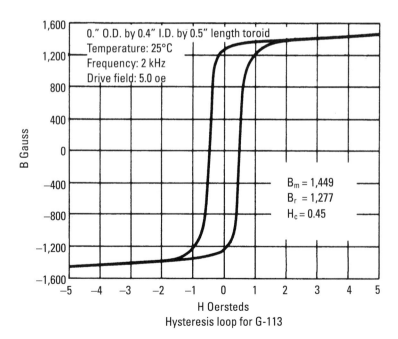

Hysteresis loop for G-113

Figure 8.7 The BH loop of a typical ferrite.

the direction of the stress [15]. Mn is used to reduce magnetostriction in YIG (Figures 8.8 and 8.9) and Al- and Gd-doped YIG. Note that YInVIG, because of its very low anisotropy, is more sensitive to magne-

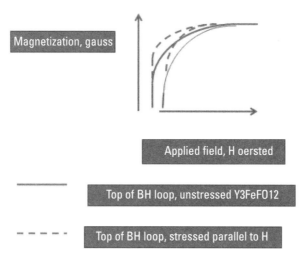

Figure 8.8   Effect of stress applied parallel to the magnetic field on YIG (negative magnetostriction in the 100 direction). (*After:* [15].)

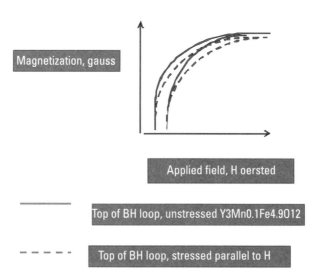

Figure 8.9   Effect of stress on Mn-substituted YIG with positive magnetostriction. (*After:* [15].)

tostriction (Figure 8.10). YZrVIG materials, which are used currently, behave in the same way.

Manganese is typically used in GdAlYIG garnets to minimize magnetostriction in rectangular toroids in WG. Because of the expected direction of stress, 0.09 Mn is used in the garnet formula (Chapter 1). CaZrVIG garnets are not used because they are not intrinsically "square."

A further consideration is the current required to switch the ferrite to a given state. This is a function of the coercive force, Hc (Figure 8.7), which is typically less than 1 Oersted for dense microwave ferrites with normal grain size. Hc is determined by magnetocrystalline anisotropy; magnetoelastic effects, a function of the ferrite chemistry; and grain size, second phase, and porosity, which are stoichiometry and processing related.

The choice of material for below-resonance switching devices is influenced by the squareness and magnetostriction requirements. As a result, the garnets in Table 8.1 are usually modified with Mn. MgMn, MgMnZn, and MgMnAl spinels have intrinsically low magnetostriction and good squareness and therefore can be used without modification. Co^{+2}, added to improve peak-power handling, will affect magnetostriction at high doping levels, however. Nickel spinels have

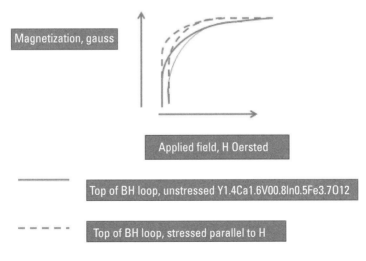

Figure 8.10   Effect of stress on YInV garnet materials. (*After:* [15].)

poor squareness and reasonably high magnetostriction. NiZn spinels improve with Zn levels if doped with Mn. However, for some applications where high Curie temperatures and good squareness and magnetostriction are required, it may be necessary to use Li spinels. The key properties of each group are summarized in Table 8.2 for S- and C-band phase shifter devices using typical MnGdAl garnets, LiTiZn spinels, and MgMnAl spinels with 4πMs in the range 440 to 680 and 1,000 gauss, respectively (adapted from [16]).

At S-band, the figure of merit expressed in degrees phase shift per decibel of insertion loss favors garnet, mainly because of lower dielectric losses. However, it can be argued that LiTiZn spinel is cheaper compared to the rare earth containing garnet. At the C-band, the figure of merit difference favoring garnet is less, with Li and MgMnAl being similar to each. Note that in this case the squareness is expressed as Br/4πMs, giving a lower value than the conventional Br/Bs. However, the Curie temperature of the LiTiZn spinel is significantly higher than the MgMnAl spinel of the same magnetization. Caution should be used in unfavorably comparing the lower Curie temperature of MnGdAl garnets with Li spinels because of the compensation effect of Gd, which does not occur in Li or Mg spinels (Chapters 1 and 2). The advantage LiTiZn has over MgMn continues up to 4πMs levels of about 2,000 gauss, but diminishes rapidly thereafter. At high 4πMs values (4,500–5,000 gauss), the NiZnMn system can be used instead of LiZn, by suitable adjustment of the squareness.

Table 8.2
Comparison of Phase Shift Ferrites at S- and C-Band

	S-Band	S-Band	S-Band	C-Band	C-Band	C-Band
Composition	MnGdAl	LiTiZn	LiTiZn+Co	MgMnAl	MnGdAl	LiTiZn
Br/4πMs	0.691	0.636	0.702	0.53	0.635	0.7
Coercive Force (Oe)	0.7	1.5	0.9	0.5	0.7	1.0
Spinwave Line Width	6.5 Oe	3.2 Oe	7.8 Oe	5.8 Oe	6.0 Oe	6.3 Oe
Degrees Phase Shift/dB Loss	650	500	400	500	550	500

Source: [16].

8.11 Temperature Considerations

One useful approach that is particularly important for phase shifters, but relevant to junction devices, is designing for performance over temperature. For all ferrites (except heavily Gd-doped materials where the magnetization can fall), both the magnetization and line width are highest at the lowest temperature of operation. This means that the insertion loss from intrinsic magnetic losses, low field loss and phase shift will be highest at the lowest temperature and therefore should meet the specification requirement for both, plus any potential nonlinearity from lack of saturation. At the highest temperature, the magnetic losses are lowest but the phase shift and spinwave related power handling are also at their lowest and therefore must be capable of making the specification.

8.12 Above-Resonance Devices

Above-resonance device material selection has very little in common with below resonance. By far the major consideration is ferrimagnetic resonance line width, either at the 3-dB or 20-dB level. This is because the ferrite is operated near resonance on the high field side. This virtually excludes all spinel ferrites and most garnets as the practical limit of use is about 50 Oersted line width at the 3-dB level for a very narrow-band device of about 1%, and values of about 20 Oersted for required bandwidths in the cellular frequencies (25 to 40 MHz at 800–900 MHz, up to 100 MHz around 2 GHz). Spinels with line widths in the 50- to 100-Oersted region have been used, but typically have very low Curie temperatures compared with corresponding garnets of the same magnetization.

The second consideration is magnetization. Because we are limited to garnets, the maximum possible magnetization is less than 2,000 gauss. In addition, because relatively strong fields are used to bias the ferrite, saturation is less of an issue, although this assumption needs to be examined closely if very low insertion loss and IMD is required. At low frequency, it cannot be assumed that the material is saturated even though it is being biased above resonance, as these are two essentially independent situations, as can be seen with single crystal YIG spheres biased to resonance when tuned below about 3.5 GHz. At about that frequency, the YIG sphere is still at resonance but begins to lose saturation, increasing losses and hence lowering the Q. For lower-frequency devices, single-crystal YGaIG with progressively lower magnetization is used. In an above resonance ferrite device, the internal field in the disk is higher because it is biased much further away from resonance. However, this simply means that the frequency below which lack of saturation becomes apparent is lower. Of course, because of the difficulty of achieving a uniform field at the edges of the disk as shown in Figure 8.4, low field loss or possibly resonance absorption will occur, and with it the possibility of nonlinear behavior in the form of IMD. Experience has shown that with polycrystalline YIG (magnetization 1,780 gauss) disks in above-resonance devices, there is a very small increase in IMD at 2 GHz, becoming increasingly worse at 900 MHz, and becoming quite severe at 400 MHz if appropriate measures are not taken. Under these conditions the saturation characteristics of the ferrite become important, such that stresses from machining become a factor. In addition to the use of precautions discussed in Section 8.5 such as multiple ferrite rings, another, simpler solution is to lower the magnetization. Under these circumstances, it may be necessary to trade bandwidth for lower IMD because of the lower magnetization. Observations that high magnetization can reduce IMD are erroneous because of other factors influencing the saturation such as size, shape, mechanical stresses, bias uniformity, and the saturation characteristics of different ferrite materials, thus obscuring the IMD effect. The most accepted general observation about IMD, which is that it will always improve with higher bias fields, tends to support this observation. This higher bias effect also occurs at high temperatures, where the saturation characteristics of ferrite degrade (measured as the BH loop squareness), even though the magnetization falls and the line width is lower, suggesting saturation is more important than the latter two.

In some cases, the size or energy product (strength) of the magnet may be an issue, and in that situation, much lower magnetizations are required, regardless of frequency, to achieve above-resonance operation, usually at the expense of bandwidth. Because above-resonance devices can be operated nearer resonance to achieve small size from a larger RF permeability, above-resonance devices have been used in cellular handsets. These have limited bandwidth and high losses as a result, but are practical for some applications. A further consideration is the magnet itself. The radial field of the magnet should be measured at the circumference nearest to the ferrite junction. This may vary unpredictably, and while it can be modified using a pole piece, should be identified and if possible specified to the magnet supplier.

Suggested materials for above-resonance applications are shown below in Tables 8.3 and 8.4. Table 8.3 allows the best choice of magnetization when magnet size, bandwidth (BW), IMD, and power

Table 8.3
Magnetization Against Frequency-Device Considerations

Magnetization Gauss	2.5–3 GHz Considerations	2–2.5 GHz Considerations	700–1,000 MHz Considerations	400 MHz Considerations
1,850	Magnet size	IMD?	IMD, power	IMD, power
1,780	Magnet size	—	IMD?	IMD, power
1,600	Magnet size?	—	IMD?	IMD, power
1,200	—	BW?	BW?	—
1,000	—	BW?	BW?	—
800	—	BW	BW	BW?

Table 8.4
Line Width Versus Temperature Stability for Various Magnetizations

Composition	Magnetization Gauss	Line Width Maximum Oersted	$4\pi Ms/$ Temperature %/ °C, RT to 85°C	Curie Temperature (°C)
YZr	1,850	10	−0.33	215
YIG	1,780	21	−0.20	280
YZrV	1,600	10	−0.29	220
YZrV	1,200	10	−0.32	208
YAlIG	1,200	25	−0.25	230
YZrV	1,000	10	−0.35	200
YAlIG	1,000	25	−0.25	210
YZrV	800	15	−0.37	190
YAlIG	800	25	−0.30	195

handling are considered at common communication frequencies. A question mark means some trade-off may be necessary. Table 8.4 gives the choice of materials when the magnetization is chosen from Table 8.3, and selection based on insertion loss versus temperature stability is required. In general, high-energy product magnets have more stable DC fields over temperature than the magnetization versus temperature slope of garnets, but hexagonal ferrite magnets more closely match those of the more stable garnets such as YIG itself. Note that none of the ferrites presented have compensation with Gd, so their magnetization slopes retain the same sign above and below room temperature. In general, YIG and YAlIG compositions are significantly more temperature stable than YZrV compositions of the same magnetization.

8.13   Power Handling in Above-Resonance Devices

This is rather like the situation with IMD. In principle, above-resonance ferrite devices will not generate spin waves and therefore nonlinear behavior should not occur. In practice, because of potential lack of saturation and the proximity to resonance, nonlinearity is possible, and similar precautions to those for IMD are necessary.

In summary, however, it should be said that the IMD and power performance of above-resonance devices is potentially better than below resonance if only because the bias fields are higher at the same frequency. However, the overlap in frequency where a choice is necessary is quite small. Above about 3 GHz, magnet size and strength limit the use of above-resonance devices. The difficulty of making a reasonably low line width garnet with sufficiently low magnetization, but a Curie temperature above 100°C, limits the use of below-resonance devices to about 1 GHz, because this represents minimum magnetizations in the range of about 250 to 300 gauss unless high insertion losses are acceptable. This is because the bias has to fall very precisely into the "valley" between low field loss and resonance on the low frequency side.

8.14   Above-Resonance Phase Shifters

Materials for this type of application [17] are very similar to those used for above-resonance junction devices, and Tables 8.3 and 8.4 can be used. Above-resonance devices cannot readily be switched because of the large DC bias on the ferrite.

8.15 Devices at Resonance

Resonance isolators are not used widely today as they have limited performance. However, they are useful waveguide devices in some circumstances. The isolation and bandwidth are directly determined by the actual resonance line width of the ferrite. Bias is typically applied with a C-shaped steel magnet. "Self-biased" hexagonal ferrites, using their large internal field, have been used up to the millimetric frequency range, but are uncommon now. YIG resonators and filters are discussed in Chapter 11 on tunable devices.

References

[1] Bosma, H., "On Stripline Y-Circulation at UHF," *IEEE Trans. MTT*, Vol. MTT-12, No. 1, 1964, pp. 61–74.

[2] Fay, C. E., and R. L. Comstock, "Operation of the Ferrite Junction Circulator," *IEEE Trans. MTT*, Vol. MTT-13, No. 1, 1965, pp. 1–13.

[3] Davies, J. B., and P. Cohen, "Theoretical Design of Symmetrical Junction Stripline Circulators," *IEEE Trans. MTT*, Vol. MTT-11, 1963, No. 12, pp. 506–512.

[4] Linkhart, D. K., *Microwave Circulator Design,* Norwood, MA: Artech House, 1989.

[5] Schloemann, E., "Microwave Behavior of Partially Magnetized Ferrites," *J. Applied Phys.,* Vol. 41, 1970, pp. 204–214.

[6] Krowne, C. M., and R. E. Neidart, "Theory and Numerical Calculations for Radically Inhomogeneous Circular Ferrite Circulators," *IEEE Trans. MTT,* Vol. 44, No. 3, 1996, pp. 419–431.

[7] Pardavi-Horvath, M., and X. M. Huang, "Local Demagnetizing Tensor Calculation for Rectangular and Cylindrical Shapes," *IEEE Mag.,* Vol. 32, No. 5, 1996, pp. 4180–4182.

[8] Pardavi-Horvath, M., et al., "Experimental Determination of an Effective Demagnetization Factor for Non-Ellipsoidal Geometries," *J. Appl. Phys.,* Vol. 79, 1996, pp. 5742–5744.

[9]   Schloemann, E., and R. E. Blight, "Broadband Stripline Circulators Based on YIG and Li-Ferrite Single Crystals," *IEEE Trans. MTT*, Vol. MTT-34, No. 12, 1986, pp. 1394–1400.

[10] How, H., et al., "Nonlinear Intermodulation Coupling in Ferrite Circulator Junctions," *IEEE Trans. MTT*, Vol. 45, No. 2, 1997, pp. 245–252.

[11] Muira, T., "Evaluation of Intermodulation Distortion of a Finite Element by the Two-Tone Method," *IEEE Trans. MTT*, Vol. 57, No. 6, 2009, pp. 1500–1507.

[12] Schloemann, E., "Integrated DC/RF Design of Dual-Ferrite Circulators," *IEEE Trans. Mag.*, Vol. 33, No. 5, 1997, pp. 3430–3432.

[13] Wu, Y. S., et al., "A Study of Non-linearities and Intermodulation Characteristics of 3-Port Distributed Circulators," *IEEE Trans. MTT*, Vol. MTT-24, No. 2, 1976, pp. 245–252.

[14] Green, J. J., and F. Sandy, "Microwave Characterization of Partially Magnetized Ferrites," *IEEE Trans. MTT*, Vol. MTT-22, No. 6, 1974, pp. 641–645.

[15] Dionne, G. F., "Temperature and Stress Sensitivities of Microwave Ferrite," *IEEE Trans. Mag.*, Vol. 8, No. 9, 1972, pp. 440–443.

[16] Baba, P. D., et al., "Fabrication and Properties of Microwave Lithium Ferrites," *IEEE Trans. Mag.*, Vol. 8, No. 1, 1972, pp. 83–94.

[17] Johnston, C. M., "Ferrite Phase Shifter for the UHF Region," *IRE Trans. MTT*, Vol. 7, No. 1, 1959, pp. 27–31.

Selected Bibliography

Cruickshank, D. B., "Understanding and Improving Insertion Loss and Inter-modulation in Microwave Ferrite Devices," *Advances in Multifunction Materials and Systems, Ceramic Transactions,* Vol. 219, New York, Wiley, 2010.

Helszajn, J., *Ferrite Phase Shifters and Control Devices,* New York: McGraw-Hill, 1989.

9

Resonators and Filters Based on Dielectrics

9.1 Introduction

There are a wide range of resonators possible using dielectrics, all of which in principle can be made into filters by coupling two or more together. The coupling is determined by the type of resonator, which can be printed on a circuit, a piece of dielectric, a dielectric/cavity combination, or a dielectric bounded by metal. This chapter will discuss resonator types and the best material for each, beginning with the lowest Q structures and moving to the highest.

9.2 Circuit-Based Resonators

These include resonators made from the type of printed circuit–based laminate materials discussed in Chapter 4 and substrates based on the ceramic-based materials covered in Chapters 5 and 6. The majority of these types of circuit are based on microstrip transmission lines. However, the simplest case is the lumped element case, valid for many circuits up to about 1 GHz when due attention is paid to the size of the element relative to the wavelength, λ (ideally < 1/100th λ). In this case the dielectric is part of a capacitor, and the angular frequency is then $(1/LC)^{1/2}$, where C is the capacitance and L is the inductance. Since the capacitance is directly proportional to the dielectric constant, which in this publication will be referred to as Er, it also follows that in this case the frequency is proportional to $(1/Er)^{1/2}$ and that the frequency drift will be proportional to $(1/Te)^{1/2}$, where Te is the temperature coefficient of dielectric constant. The Qs obtained with lumped element designs are quite modest, depending on dimensional factors, but a good rule of thumb is that 1/loss tangent of the dielectric should exceed the desired total Q by at least a factor of 10 at the desired frequency. Hence a desired Q of 100 will require a loss tangent of 0.001 from the material at the desired frequency, a modest value that allows the use of a very wide range of materials. For most circuits, a Te of +/− 10 ppm/°C is adequate for this level of Q. If the dielectric's temperature characteristics are specified as Tf, the temperature coefficient of frequency from a fully loaded dielectric loaded transmission line measurement, the Te can be derived from Tf = $-(1/2Te + \alpha_1)$, where α_1 is the thermal expansion in ppm/°C.

Transmission line resonators in a fully filled dielectric medium consisting of a half wavelength–long resonator also have a frequency dependence proportional to $(1/Er)^{1/2}$ and the frequency drift expression above also applies, for example in a dielectric-filled coaxial TEM resonator. This means that for a Tf of 0, the Te must be $-2\alpha_1$, or about −20 ppm/°C for most ceramic dielectrics. For plastics and most plastic/ceramic composites, the expansion coefficient is much higher and a reasonably stable Tf is not possible, especially since most plastic and ceramic materials' dielectric constant falls with increasing temperature. One exception is silica- and glass-based laminates in the XY direction, where the very low expansion fiber dominates over the plastic, and a printed resonator can be made reasonably stable. In the Z-direction, that is the thickness; the plastic dominates the expansion, a factor which influences the behavior of microstrip or triplate in these types of filled plastic. Fully filled triplate transmission lines are formed by fusing Teflon or Teflon/ceramic composite glass laminates together.

9.3 Coaxial Resonators

Coaxial resonators are an example of another fully filled transmission line, usually as a quarter-wave resonator. The choice of material is determined by a number of factors. The size of a resonator's cross-section is determined by the required impedance and by the maximum height allowed, usually determined by the sum of the PCB thickness and the surface-mounted resonator. The length of a quarter-wave resonator of course is $\lambda/4 \times (1/Er)^{1/2}$, where λ is the wavelength for the desired frequency. However, to avoid other TEM modes, the length should ideally be at least twice the cross-section height (square cross-sections are used for ease of mounting). Since most cross-sections are 2 to 6 mm (0.08 to 0.024 inch) in height, to maintain the length at a high frequency a low dielectric constant (about 10 at 6 GHz) is necessary; also, to prevent the length becoming too great at a low frequency, a high dielectric constant (about 90) is used. This effect can be seen in Figure 9.1, comparing the quarter-wave resonator length, L, in mm, for dielectric constants of 81, 36, and 9 over the frequency range 300 MHz to 6 GHz (1 inch = 25.4 mm). For resonators in the range of 2 to 6 mm cross-section, typical examples are at frequencies from 600 MHz

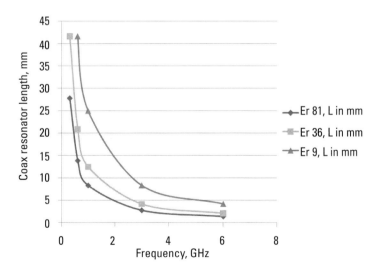

Figure 9.1 Coaxial resonator length (mm) versus frequency (X-axis, GHz) for different dielectric constants.

(Er 81, 6 × 6 mm cross-section, 12-mm length) to 6 GHz (Er 9, 2 × 2 mm cross-section, 4-mm length)

The Q possible with quarter-wave coaxial resonators from two materials (Er 20 and 90) is shown in Figure 9.2 for a range of cross-sections from 2 mm to 12 mm. The Q is largely a function of size, because of the conductivity of the metal, in this case thick-film silver, making up the center and outer conductor and the short circuit end. However, the much higher loss tangent of the Er 90 material also influences the Q. The Q calculation is conservative, as actual values are 10 to 15% higher. Extrapolation to other frequencies is possible assuming the Q increases as the square root of the frequency.

By using coaxial resonators with a 25 × 25-mm cross-section, it is possible to achieve Qs in the region of 1,500 at 1 GHz with low dielectric loss materials. If a larger size is permitted by the application, other resonator approaches using TE or TM modes should be considered.

Typical materials that are actually used for coaxial resonators are shown in Table 9.1. Note that the same type of materials can be used as capacitors for coupling structures for coaxial filters and as substrates for printed resonators. However, account must be taken of the different temperature coefficients for capacitors (Te); for fully filled structures like coaxial resonators, Tf; and, as we will see, partially filled structures like dielectric resonator/cavity combinations where the Tf is more complex in origin. Alumina is included as it is sometimes used to make

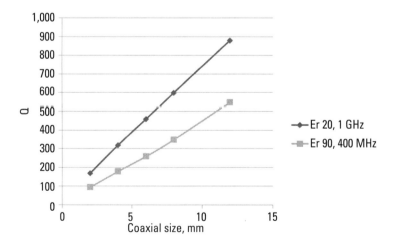

Figure 9.2   Q versus coaxial size in mm (X-axis) for dielectric constants 20 and 90.

Table 9.1

Materials for Coaxial Resonators and Dielectric Capacitors and Substrates

Material	Source	Er	QF Product	Te (ppm/°C)	Thermal Expansion (ppm/°C)	Tf (ppm/°C)
D1000	TT	10.5	> 20 THz	−15	7.5	0
Alumina 99.5	CoorsTek	9.6	100 THz	+94	6.0	−53
D2000	TT	20.6	> 20 THz	−17	8.5	0
D8800	TT	39	> 20 THz	−25	9.5	+4
D9000	TT	90	> 5 THz	Variable	Variable	0

capacitors on circuits, but is not temperature stable enough for resonators. Note that in Table 9.1 the Tf is calculated from the Te and expansion coefficient, and does not refer to the Tf in manufacturers' literature, which is derived from their standard cavity data (Chapter 12).

Note that Er 90 materials are TTBs (Chapter 6) and are anisotropic. In practice, different compositions are necessary to achieve near zero Te for capacitors and near zero Tf for fully filled structures like coaxial resonators. Substrates manufactured by die pressing have the plate-like structure perpendicular to the direction of pressing, giving an anisotropic dielectric constant and temperature coefficient. Coaxial resonators made by die pressing behave similarly. When coaxial resonators are made by extrusion, the plate-like structure tends to align with the extrusion direction, giving different values, and again the composition must be changed to give a near-zero Tf. The Tf value in Table 9.1 is for an extruded 4-mm coaxial resonator of a particular composition (TT = Trans Tech Inc.). The same composition cannot be used to give a zero Te for a die-pressed substrate, or a zero Tf for a die-pressed coaxial resonator, although both options are available as different compositions.

9.4 TE-Based Dielectric Resonator Applications

Dielectric resonator-based oscillators, mounted onto a microstrip dielectric substrate with an adjustable height metal housing above it, have been studied by Higashi and Makino [1] and developed by Ishihara [2]. Essentially, because the frequency change with temperature is a function of the dielectric constant coefficients of the resonator and substrate and the expansion coefficients of the latter two plus the housing, stability is only possible over a narrow frequency range, but is possible over a wide temperature range for a given height of the housing with a fixed combination of the two dielectrics. By changing the height (the equivalent of using a large metal tuning screw), it is possible to find other narrowbands of operation with good frequency stability.

In the case of a dielectric resonator being used as an oscillator element, mounted on a support in proximity to a microstrip line without a metal housing, temperature-stable operation (that is Tf = 0) is proportional to the inverse of the loaded Q (adjustable by moving the resonator relative to the microstrip line or by adjusting the support height) and the temperature differential of the phase angle that sets the oscillator frequency [3]. Because it is difficult to relate these coefficients to the published Te or Tf of the material, in practice it is necessary to optimize the oscillator performance first by adjusting the coupling, then substituting resonators of different Tf or Te until temperature stability is achieved. Therefore, in general, the temperature coefficient of the support dielectric is also a factor. Ideally the support dielectric constant should be as low as possible, usually 4 to 6, with QF products > 20 THz. Plastic supports are not recommended because of their large expansion coefficient. Materials used in dielectric resonator oscillators have both size and Q requirements broadly similar to cavity-based dielectric resonators and filters, but are typically used over much wider frequency ranges. For an isolated resonator, the frequency is proportional to $1/Er^{1/2}(D^2L)^{1/3}$. When the diameter D is equal to twice the thickness, L, this becomes proportional to $1/Er^{1/2}(D)$. Using the correct proportionality constant for an isolated puck resonator, this becomes frequency (GHz) = $296.66/Er^{1/2}D$, where D is the diameter in mm. It can be seen from Figure 9.3 that a high dielectric constant is necessary at low frequencies to obtain a reasonably small resonator, and from Figure 9.4 a low dielectric constant is necessary at high frequencies to obtain a part that is easier to machine to tight tolerances because it is larger. These frequency and size calculations are based on an isolated resonator and are for guidance only.

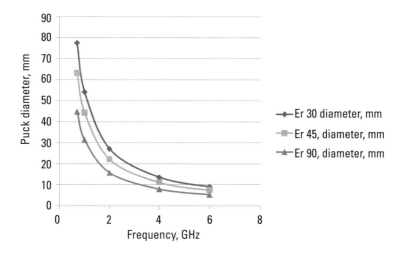

Figure 9.3 $TE_{01\delta}$ puck diameter (mm) versus frequency, (X-axis, 700 MHz to 6 GHz).

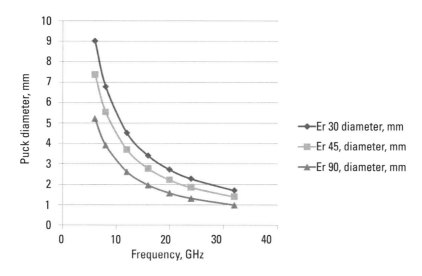

Figure 9.4 $TE_{01\delta}$ puck diameter (mm) versus frequency, (X-axis, 6 GHz to 32 GHz).

9.5 Dielectric Resonator Loaded Cavities

Figures 9.3 and 9.4 also apply approximately to $TE_{01\delta}$ resonators in cavities (Figure 9.5). In particular, the diameter will change signifi-

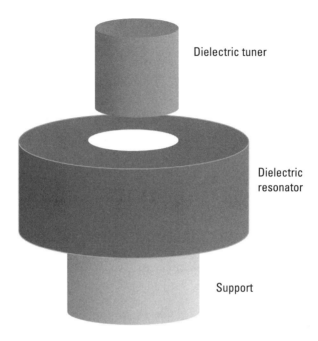

Figure 9.5 Dielectric resonator puck and tuner with support for cavity mounting (cavity omitted).

cantly if the thickness does not equal half the diameter. Note also cavity wall effects are ignored, and the diameters given are for guidance only. Since the cavity dimensions have to substantially exceed the resonator diameter and thickness to achieve the necessary Q (Figure 5.1), it follows this also affects the total cavity volume.

The resonator volume, in this case $\pi.D^3/8$ cubic mm, in turn adds substantial weight at low frequencies, given densities in the range 4.5 to 6.5 g/cc. As a result a 700-MHz resonator with a 60-mm diameter may weigh around 450g with an Er of 45 and a volume of 85 cc. The same resonator with an Er of 90 will be less than half the Er 45 weight, around 200g, making a substantial difference to total weight in a multiresonator filter mounted on the top of a cell phone infrastructure tower. Currently, however, no materials exist with an Er of 90 and a comparable QF product to available Er 45 materials (Table 9.2).

The temperature coefficient of frequency, Tf, for a partially dielectric-filled cavity is difficult to calculate because of the contribution from the cavity and dielectric support and the difficulty of analyzing an

Table 9.2
Dielectrics for TE Mode Applications

Material/ Source	Application (TE Mode)	Er	QF Product	Tf (ppm/°C)	Relative Cost
BMW Kyocera	2 GHz? > 10 GHz	21	150 THz	−2 to +4	Contains Tungsten
BMT NTK	> 10 GHz	24	200 THz	−2	Contains tantalum
BZT various	2 GHz? > 10 GHz	29–31	110 THz	+5 to −5	Contains tantalum
BCZN various	2 GHz > 5 GHz	33–35	90 THz	+5 to −5	Contains niobium
Titanate based various	900 MHz to 20 GHz	35–39	40–65 THz	+10 to −10	Low or no expensive elements
NdAl/CaTi Kyocera	700 MHz to 20 GHz	45	40 THz	−3 to +8	Contains RE
ZrTitanate + Nb, Zn, various	700 MHz to 20 GHz	43	40 THz	−3 to +9	Contains niobium
TTBs various	700–900 MHz	75–90	5–12 THz		Contains various REs

inhomogeneously dielectric-filled cavity. In practice, computer simulations based on Maxwell's equations are used to solve for the size of the cavity and resonator to achieve the necessary frequency and Q for a required maximum cavity size, if necessary using empirically derived conductivity values for the type of cavity plating being used. At that point, the necessary temperature drift of the resonator/cavity combination must be established iteratively, by substituting materials of the approximately same dielectric constant with different Tfs established from independent standard cavities (Chapter 12). It follows that each type of dielectric must be available in a wide range of Tfs to cover all possible cavity sizes for a given frequency.

Tuning of cavity-based dielectric resonators by mechanical methods is done either with large metal screw tuners, typically the same diameter of the puck, or by dielectric tuners, made of the same or similar material as the resonator. There are two main types of dielectric tuner: a disk type and a plunger type (Figure 9.5). The disk type is usually the same diameter as the puck, and sufficiently thick to effectively increase the thickness of the puck, driving the frequency down (note that a metal tuner has the opposite effect). However, as can be expected, the dielectric tuner is the most effect near the puck and therefore the tuning characteristic is not linear with distance from the top face of the puck. As a result, this method is not usually practical for more that about 1% of frequency tuning, and does not lend itself to motorized tuning, where linearity is usually required. A well-used alternative is the plunger type, where the resonator has a through hole just large enough to allow the plunger cylinder through. In this situation, as the plunger bottom moves from just above the resonator through to the bottom of the resonator hole, there is a sufficient region of linearity, typically at least 3% of the frequency. This principle is used in both motorized autotuned combiner resonator cavities and mechanically tuned filter cavities. The materials selected for resonator and tuner appear in Table 9.2. As some high-Q resonator materials are expensive because of their raw material costs, the disk tuner is often made of a cheaper material as the net Q of the resonator/tuner assembly does not degrade much. This is not usually possible for the plunger-type tuner as Q normally has to be maintained at the deepest level of insertion. Although autotuned combiners are gradually being replaced in communication systems, there is currently a requirement for autotuned filters to allow tuning electronically either locally at a base station or remotely through the system control network. These may not require the tuning bandwidth of an autotuner combiner but require reasonable linearity because of the complexity of tuning a multicavity filter, as all filter attributes (return loss, insertion loss, attenuation, passband bandwidth, and so forth) must be maintained. As a result, tuner plungers, which have either stepped or tapered cross-sections, are used, usually made from the same dielectric as the resonator. Table 9.2 lists the broad types of dielectric materials available. Because of cost considerations, those containing significant amounts of tungsten, tantalum, niobium, and rare earths (REs) are highlighted. Currently, tantalum-based materials are only used at a high frequency (Figure 9.4) because of cost. Niobium is much less expensive, but still significantly more expensive than titanium, and therefore materials with significant amounts of niobium are noted. Rare earth materials such as samarium and neodymium have become significantly more expensive currently, a situation that seems likely to last for several years, and it is not expected their costs will revert to those of the 1990–2010 era.

From Table 9.2 it can be seen that the primary application for low dielectric constant, high cost materials is very high Q, high frequency dielectric resonators, mainly for oscillators in the range 10 to 40 or more GHz. Titanate-based materials low in expensive elements are useful as high-Q resonators over a wide range of frequencies and are also suitable for low-cost tuners. Niobium-containing materials may be expensive, depending on the amount present. Those containing rare earths (REs) can be expected to become much more expensive, and may not be used for TE applications in the future where large pucks are required.

9.6 Dielectric Support Materials

Support materials ideally have a very low dielectric constant, high Q, and reasonably low Tf or Te. Candidate materials are discussed in Chapter 5 in depth. For TE ceramic-based transmitter filters and combiners, an additional requirement is high thermal conductivity, which tends to leave only one material of choice, namely alumina, as the best means of heat sinking the resonator to the metal of the cavity. Alumina has quite a high dielectric constant and is not very temperature stable, which, combined with the cavity metal expansion effects, requires compensation of the overall frequency drift by adjusting the resonator Tf. However, no comparable material is available (Tables 5.1 and 5.2), although lower conductivity, lower dielectric constant materials such as cordierite, steatite, and fosterite, are suitable for low-power applications. In general, plastics are not suitable because of their high thermal expansion, but glass-filled plastics such as Ultem are used because of their improved mechanical stability.

Ultem has quite a high dielectric loss (Chapter 4) and must be used sparingly to avoid significant loss of Q. Lower loss glass-filled materials are less stable thermally, but are useful in some applications. Examples are when the alumina support causes spurious modes to appear in-band in cavity resonators or filters. If the volume of the alumina cannot be changed (depending on which side of the frequency the spurious response appears) a lower dielectric constant support may be a possible solution.

Gluing of the support to the resonator will reduce Q because of the dielectric loss of the glue. This can be minimized by using alumina-filled silicones as glue. Although not giving a particularly strong bond, their flexibility takes up thermal expansion mismatches between materials, one of the prime causes of glue joint failure.

9.7 TM Dielectric Resonator-Based Cavities

Filters whose specification allows dielectric-loaded cavity solutions with Qs in the 3,000 to 10,000 region can use TM mode structures, consisting of a cavity with a rod or tube of dielectric. The size of a TM mode rod can be much less than the corresponding TE mode resonator, but at the expense of Q. This can be seen in Figure 9.6.

From Figure 9.6 it can be seen that to cover the frequency band of 2 to 3 GHz, dielectric constants in the range 20 to 35 are useful for rods in the 8- to 12-mm diameter range. The frequency can also be raised by having a reduced height rod relative to the cavity height. At frequencies

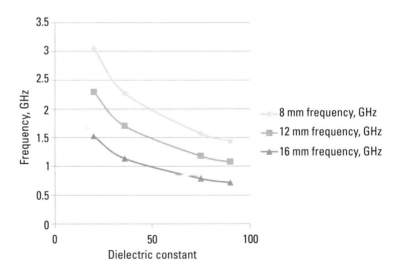

Figure 9.6 Approximate dielectric constant required versus frequency (Y-axis, GHz) for a full height (38 mm) TM rod in a 55-mm diameter cavity, with rod diameters of 8, 12, and 16 mm.

in the 700- to 900-MHz range, the dielectric constant needs to be in the range 75 to 90 to keep the rod diameter below 16 mm. Note that only the lower frequency bands are fully tested or simulated data; the upper bands are extrapolated data, assuming the frequency is proportional to $Er^{1/2}$ and inversely proportional to diameter, and are subject to error. The measured Qs with 38-mm full height rods of a dielectric constant of 75 to 80 and a QF product of 12 THz produce cavity Qs in the region of 5,000 to 6,000 when the cavity diameter is 55 mm.

The main practical issue with TM rods is fixing the rod to the cavity bottom and, if full height, also to the top, because of the thermal expansion mismatch between the metal cavity. This can be done by thick-film metallization of the end(s) of the rod using a high conductivity silver silk screen paste, compatible with the rod dielectric to ensure good adhesion, then soldering or brazing the rod or tube into the cavity either directly or indirectly. Alternatively, partial height tubes with ceramic or plastic inserts drilled and tapped to allow screw fixing can be used, provided the screw is suitably plated and does not protrude significantly into the cavity. Mechanical methods using flexible conductive diaphragms for the cavity top are also possible.

9.8 Intermodulation in Dielectric Loaded Cavities

The contribution from the cavity components can be broken down into these created by dissimilar metals in contact, and those from the dielectric materials used in the cavity. Normal practice is to use silver electroplating on all metal surfaces in the cavity, including internal fixing screws and tuning screws, such that silver to silver contact is maintained. Partial plating of cavities to reduce silver costs must take that into account. Copper plating is an option provided the copper does not oxidize, but silver is less likely to lose Q quickly through atmospheric corrosion, and is generally preferred. Screw threads are particularly important because of the difficulty of plating the inner part, and wear on the thread, which may reduce the thickness of the silver or expose the base metal. Note also that screws made of poorly conducting metals, such as stainless steel, should be plated with silver to prevent loss of Q, regardless of intermodulation considerations.

Intermodulation from dielectric materials in resonators has been studied by Tamura [4]. It was concluded that third-order intermodulation distortion was broadly a function of the dielectric loss of the dielectric, although some materials of equivalent loss showed different intermodulation, and this was thought to be a function of individual materials' crystal structure. High purity dielectrics were significantly better, in line with their loss tangents. A TTB material, barium lead neodymium titanate–based, showed poor intermodulation. Although this could be linked simply to the low Q of the TTB material, it may also relate to the tendency of this class of materials to be ferroelectric in related compositions, indicating an incipient ferroelectric type of nonlinear relaxation might be possible (Chapter 6). The use of TTBs in situations where low intermodulation is required should certainly be approached with caution.

References

[1] Higashi, T., and T. Makino, "Resonance Frequency Stability of the Dielectric Resonator on a Dielectric Substrate," *IEEE Trans. MTT*, Vol. MTT-29, No. 10, 1981, pp. 1048–1052.

[2] Ishihara, O., et al., "A Highly Stabilized GaAs FET Oscillator Using a Dielectric Resonator Feedback Circuit at 9-14 GHz," *IEEE Trans. MTT*, Vol. MTT-28, No. 8, 1980, pp. 817–824.

[3] Tsironis, Y., and V. Pauker, "Temperature Stabilization of GaAs MESFET Oscillators Using Dielectric Resonators," *IEEE Trans. MTT*, Vol. MTT-31, No. 3, 1983, pp. 312–314.

[4] Tamura, H., et al., "Third Harmonic Distortion of Dielectric Resonator Materials," *Jap. J. Appl. Phys.*, Vol. 28, No. 12, 1989, pp. 2528–2531.

Selected Bibliography

Kajfez, D., and P. Guillon, *Dielectric Resonators*, Dedham, MA: Artech House, 1986.

TransTech Applications Notes 821, 851, 1030, 1008, 1009, 1010, 1014, 1015, 1017, 1021, and 1030.

10

Antennas and Radomes

10.1 Introduction

Antennas cover a very wide range of technologies and architecture. The purpose of this chapter is to attempt to cover those antennas with high materials content, and where the choice of material makes a significant difference to the antenna performance. It does not specifically include metamaterials, but some of the materials likely to be used in such applications as the technology develops can be found in individual materials chapters. For convenience, radome materials have been included in this chapter.

10.2 Ferrite Rod Antennas for VHF and UHF

These types of antennas cover the range from below 1 MHz to about 200 MHz, but only those useful above 10 MHz will be considered in this chapter. Originally, the Nickel Zinc spinel series was used for these applications, usually with porous, fine-grained material to extend the frequency range. Cobalt was then used to extend the frequency of nickel spinel further. More recently, because of the emergence of the 13.56 MHz industrial, medical, and scientific band, these materials have been given more attention (Chapter 2) by fine tuning the permeability and magnetic loss of the NiZn spinels with cobalt. By bringing the relaxation peak associated with the Co^{+2} substitution and that associated with the NiZn ratio into near coincidence, the permeability can be maximized and the magnetic loss minimized, such that permeability in excess of 100 is achieved with Qs of the same order at 13.56 MHz. The same technique can be used to produce a series of NiZn plus Co materials with reducing Zn covering up to 200 MHz with permeability in excess of 10 and good Q. Fine tuning of the composition is only possible through advanced process control using high-resolution X-ray fluorescence. The spectrum of a nickel-only spinel with cobalt doping is shown in Figure 10.1, showing the permeability (about 15) to be greater than the dielectric constant of 13. Spinels for rod antennas typically have dielectric loss tangents less than 0.0005, much less than the corresponding magnetic loss tangents, which are in the region of 0.01 to 0.005.

The spectrum of a modified hexagonal ferrite (Figure 10.2) is shown for comparison as hexagonal ferrite can be used at higher frequencies. In the figure the permeability is represented by $\mu r'$, and the units as U. The magnetic loss is represented by $\mu r''$, and units in mU, equal to $\mu r''/1,000$. The magnetic loss tangent is then $\mu r''/\mu r'$,

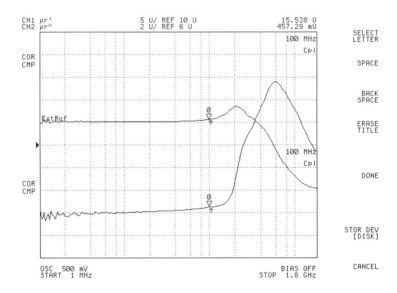

Figure 10.1   Spectrum of nickel ferrite with cobalt doping.

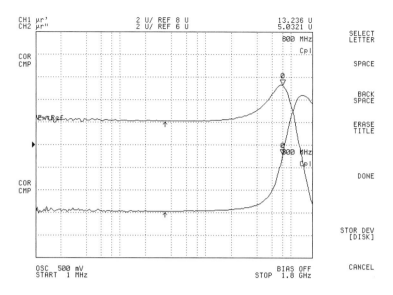

Figure 10.2   Spectrum of a modified cobalt Z-type hexagonal ferrite.

equal to 1/magnetic Q. The dielectric constant and loss are essentially independent of frequency from 10 MHz to millimetric frequencies.

The hexagonal ferrite can be seen to have a constant permeability and loss up to about 400 MHz, although the magnetic loss is somewhat higher.

10.3   Patch Antennas

These are used at UHF frequencies and above. Because they are essentially a half wave resonator, the size is proportional to wavelength and hence to the square root of the dielectric constant of the material the patch is printed on. The same range and type of ceramic materials is used for these as is for coaxial resonators, that is a dielectric constant of 10 to about 40 (Table 9.1), but with some additional considerations for temperature control for UHF and low microwave frequencies. Above an Er of 40 the bandwidth becomes extremely narrow and the impedance match to free space is very poor. Dielectric losses can be modest (less than 0.001 is sufficient) and a temperature coefficient of the order of $+/-10$ ppm/°C is adequate for most applications. However, some applications require very wide temperature ranges and more specialized materials may be necessary (Table 10.1). The most critical parameter for patch antennas is the dielectric constant, which must be controlled to within $+/-0.2\%$ within a ceramic batch to achieve a frequency accuracy of $+/-0.1\%$. Because 1.5-GHz GPS patch antennas (Figure 10.3) are narrowband, it is usually necessary to have a "library" of patch sizes to account for batch to batch variation in a dielectric constant, in the form of a series of silk screens used to print the thick-film silver ink patch on the ceramic substrate. At GPS frequencies the desirable

Table 10.1
Materials for GPS Patch Antennas

Material	Source	Er	QF Product	Tf (ppm/°C)	T'f (ppm/°C/°C)	Te (ppm/°C)	Exp Coefficient (ppm/°C)
D.1000	TT	10.5 +/−0.5	> 20 THz	0 +/− 10		−15	7.5
D.2000	TT	20.6 +/− 1	> 20 THz	0 +/− 10	−0.09	−17	8.5
D.8371	TT	35.5 +/− 1	> 40 THz	0 +/− 1	< 0.005	−20	10.0
D.8812	TT	39 +/− 1.5	> 20 THz	4 +/− 2	<0.01	−25	9.5
D.4300	TT	43 +/− 0.75	> 40 THz	0 +/− 1	−0.01	−13	6.5

Figure 10.3   GPS patch antenna used for navigation.

accuracy corresponds to $+/-1.5$ MHz. Temperature drift of $+/-10$ ppm/°C over a 50°C range will add another $+/-0.75$ MHz, so library screens may be as little as 1-MHz apart. Alumina and many other low dielectric constant materials (Chapter 5) are not generally suitable for GPS patch antennas because of the temperature requirement. In Table 10.1, the Er variation from batch to batch is given, along with the tolerance on Tf (coefficient of frequency drift with temperature), and the second-order temperature coefficient, for ceramic dielectrics used at GPS frequencies.

For higher microwave frequencies, microwave grade glass-filled laminates with dielectric constants of 10 or less are used, as listed in Table 4.8, with correspondingly lower frequency stability, in many cases depending on the expansion coefficient in the XY plane of the laminate. Foam-based dielectric with dielectric constants close to unity can also be used for high-frequency patch antennas provided moisture absorption is avoided (see Table 10.2), using inverted structures to avoid having to metalize the foam itself, which can be an issue in processing and use because of porosity and moisture absorption. Cross-linked polyolefin laminates, with low dielectric constant and low loss, can also be used (Section 4.2.2).

The efficiency, gain, and bandwidth of patch antennas can be improved by the use of a reactive impedance surface, printed on a dielectric substrate [1]. In this case a forsterite (Er ~6) patch was used, with a square metal array printed on a magnesium calcium titanate (Er 25) substrate.

10.4 Ferrite Patch Antennas

Because the permeability of ferrite can contribute to the size of a patch antenna, where the wave length is reduced by $1/(\text{Er}\,\mu)^{1/2}$, it is possible to use materials where the size reduction factor is more than 10, using spinels up to 200 MHz, and hexagonal ferrites up to at least 500 MHz, as the latter's permeability can exceed 10 over a wide frequency range. Because Er and μ are almost equal, from Maxwell's equation a good match into free space is possible over a wide band of frequencies [2]. The spectrum of a hexagonal ferrite (Figure 10.2) can be compared with Figure 10.1 for a nickel ferrite, showing the higher frequency potential of hexagonal ferrite. Currently research is being carried out to extend that range further for antennas and related applications such as high-frequency inductors. At higher frequencies, tunable ferrite patch antennas are possible (Chapter 11).

10.5   Planar Inverted-F Antennas (PIFA)

These can be seen as a special case of patch antennas where the printed resonator is essentially a function of wavelength. PIFA performance can be enhanced using high dielectric constant materials as the resonator, which can be multiband by appropriate choices of dimensions to give a series of resonant frequencies. Note, however, that the ground plane dimensions remain large relative to frequency for maximum efficiency and the total area occupied still remains relatively large.

10.6   Dielectric Resonator Antennas

This may be in the form of a puck or rectangular shape, or as an LTCC structure. Coupling is possible by slots under the resonator, and multi-resonator arrays are possible (see the Selected Bibliography). The material requirements are similar to resonators for filters, but the Q and temperature drift requirements are less severe, so a wide range of ceramic materials with an Er of 20–90 are usable, the emphasis being on cost, such that titanate-based dielectrics are invariably used (Chapter 6).

A special case of dielectric resonators is that based on multifilar spirals around high-dielectric materials, developed by Sarantel Ltd. [3]. These have superior isolation, coverage pattern, gain, and directivity to high dielectric constant patch or resonator antennas. These use high-Q (QF > 10 THz) dielectrics in the range 20 to 80.

Dielectric waveguides [4] with periodic radiating elements can also be used as antennas. Metallized plastic or ceramic dielectric waveguides can be used with suitable coupling slots to control frequency, beam pattern, and polarization. Alternatively, a high dielectric constant waveguide without metal walls can be used as a leaky waveguide.

10.7 Metal Antennas

Metal has traditionally been used for reflectors and dipole elements for antennas. The prime requirement is for electrical conductivity and light weight, so mesh structures and tubular elements are used, typically in aluminum. Environmental protection is required if the structure is not enclosed by a radome, so anodized or polymer-coated aluminum is used. Anodizing, which essentially adds a hydrated aluminum-oxide coating, is insulating, so electrical contact must be maintained between elements where required. The same is true of gradual oxidation of aluminum exposed to the environment. Skin depth is only an issue at very high frequencies and thick anodized or oxidized coatings, but significant degradation of antenna performance is possible under these conditions.

Alternatives to aluminum are metalized plastics (Chapter 7), which can be lighter and potentially more mechanically stable.

10.8 Radomes

Radomes vary from very simple structures for commercial purposes to very complex structures for military or space applications. The main requirement for a radome is transparency to microwaves, implying reasonably low dielectric constants to reduce reflections, and low dielectric losses to maximize transmission through the radome material. Distortion of the beam pattern of transmitted or received signals is also an issue relating to the thickness and shape of the radome, which may be a function of the ease of manufacture of the material being used and its structure. The angle of incidence may also be oblique, increasing the effective path through the dielectric. The type of material may also be dictated by its environment, which may vary from inside the home to the most adverse climates in the world. Special consideration may also be required for high-velocity projectiles or space environments. As a result, radomes are made of all classes of dielectric materials: plastics, plastic foams, plastic/glass composite laminates, glass-ceramic composites, and ceramic. In general, low-dielectric foam helps to reduce effective electrical thickness and physical weight.

Radome construction falls into four basic types: electrically thin, half wavelength, A-sandwich, and C-sandwich. A fifth type, the B-sandwich, is less common.

Electrically thin radomes have their thickness less than one tenth of the wavelength in the radome dielectric at their operating frequency. For broadband operation this would apply at the highest frequency being used. This type of structure includes solid plastics and glass re-inforced plastics and could include low dielectric constant ceramics at sufficiently low frequency if sufficiently robust. Relatively thick foam structures with very low dielectric constants are also possible, provided they are sufficiently sealed on their surfaces. This means that for a solid plastic with dielectric constant of 2.2, the maximum thickness would be about 2 mm at 10 GHz, if the maximum thickness allowed was one tenth the wavelength, making such materials suitable for modest environments. However, a radome made of solid alumina with dielec-tric constant 9–10 would have to be very thin for a ceramic, about 1 mm, at 10 GHz, which would make it very fragile in a severe environ-ment, virtually excluding it from use at these frequencies. The choice of basic plastics includes low-loss materials (Tables 4.1 and 4.2), high-temperature plastics (Table 4.3), and low-cost plastics (Table 4.4). The lower-cost plastics include filled types like ordinary all-purpose fiber-glass. For high-temperature operation in radomes, some of the higher-temperature plastic laminates types were measured by Petri (Chapter 4, [12]) up to 260°C at 9.35 GHz when used with glass fiber and quartz reinforcement.

10.8.1 Half-Wave Radomes

These will only pass frequencies effectively when the frequency is one half of the wavelength in the dielectric. Although a much narrower band than an electrically thin radome, this type of radome allows the use of higher dielectric constants (see Figure 10.4). It also allows greater thickness of material, and hence greater strength.

From Figure 10.4, it should be noted that for the 2-mm thin radome dimension this is the maximum frequency for each dielec-tric constant. For the half wavelength, the frequency in the plot is the center frequency, with a corresponding limited bandwidth. It will be

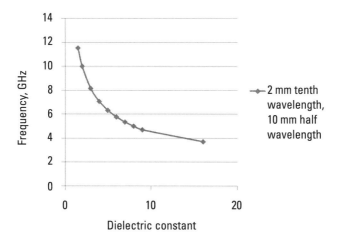

Figure 10.4 Dielectric constant (X-axis) versus frequency (Y-axis, GHz) for half-wave 10-mm and tenth-wave 2-mm material thickness.

evident that for a half-wavelength 2-mm thickness material, that is one fifth of the thickness of the half wave 10-mm plot, the frequency of use can be extended from a maximum of 10 GHz to a center frequency of 50 GHz with a dielectric constant of 2, and so on.

10.8.2 A- and C-Sandwich Construction

A-sandwich construction consists of two thin solid dielectric walls separated by a low dielectric constant foam or honeycomb structure. In theory, any combination of solid dielectric and foam is possible, but common combinations are fiberglass with plastic foams or honeycombs, and ceramic or ceramic composite materials with ceramic foam. C-sandwich construction consists of two outer solid walls, a third internal solid wall, and two cavities filled with foam or honeycomb materials.

10.9 Foam Radome Materials

Foam materials with very low dielectric constants (1 to 2) are made either by foaming agents with polymers or by enclosing hollow spheres, usually of glass, with a plastic matrix. These are usually low temperature materials but high temperature plastics such as polyimide [5] have been used. These materials can either be made by machining blocks of factory prepared materials, or by foam in place of materials, which, as the name implies, are used in the assembly and construction of antennas in situ. Foams are rarely used on outside surfaces of antennas because of their high moisture absorption and poor resistance to rain erosion and adverse climatic conditions. Examples of commercially available foams from Emerson and Cuming [6] are summarized in Table 10.2. Similar materials are available from ARC Technologies and Cuming Microwave [7, 8].

For comparison, high-temperature polyimide foams have been made with improved temperature properties, in excess of 177°C [5].

10.10   Ceramic Materials

Ceramic materials used in solid form are shown in Table 10.3, mainly for high-temperature defense or space applications use. These materials are subject to high temperatures (up to 1,200°C or more), thermal shock, and mechanical shock, plus must meet standard antenna requirements like low dielectric constant, relatively low dielectric loss , and light weight. These materials are essentially glass or glass ceramic

Table 10.2
Low Dielectric Constant Antenna Materials

Material	Form	Er	Loss Tangent	Max. Cont. Use Temperature (°C)	24-Hour Water Absorption (%)
Polyurethane	Machinable block or pour in place	1.04–1.25 depending on density (1 MHz)	0.001–0.005 depending on density (1 MHz)	135	3–1% depending on density
Thermoset	Block	1.7	> 0.0004	150	0.1
Cross-linked hydrocarbon	Closed cell Foam block	1.03 or 1.06	0.0001	85	0.04
Epoxy	Pourable	1.25 (8.6 GHz)	0.005 (8.6 GHz)	175	Porous

Table 10.3
Properties of High-Temperature Radome Materials

Property	Fused Silica	Pyroceram 9606	S-994/AlPO$_3$	Reaction Bonded Silicon Nitride	Silicon Nitride	IRBAS	Bonded Silicon
Material Source	Corning	Corning	Brunswick	Ceralloy 147-01		LM	Ceralloy 147-31N
Composition	SiO$_2$	Glass/ceramic composite	Si$_3$N$_4$ + AlPO$_3$	Variable porosity Si$_3$N$_4$	Pure dense Si$_3$N$_4$	Ba, Al-silicate + Si$_3$N$_4$	As IRBAS
Density, g/cc	2.2	2.6	1.9	1.8–2.5	3.29	3.18	3.21
High temperature strength, MP		89 at 815°C		230 at 1,400°C		502 at 704°C	365 at 1,400°C
Maximum use point (°C)	1,100	1,350	1,110	1,400	1,000	1,000	1,200
Thermal expansion ppm/°C	0.7	5.7 RT to 300°C	3.6 XY 1.1Z	3.1	3.3	3.2	
Thermal conductivity W/m.K	0.8	3.3		~6 (Var.)	30	20	25
Dielectric constant	3.78	5.5 at 8.6 GHz	3.45	4–6	7.5	7.6	8
Dielectric loss	0.0001	0.0003 at 8.6 GHz	0.008	0.002–0.005		0.002	0.002

composites with various glass to ceramic bonding agents to improve strength and thermal properties, and are completely different from more conventional high-temperature oxide ceramics, which are sometimes used, such as alumina or beryllia (Chapter 5). The most recent are composites of silicon nitride (Si$_3$N$_4$) because of its high temperature properties (summarized in Table 10.2, and also discussed in Chapter 5). Fused silica (which has been used extensively as a high-temperature radome material directly) and silicon nitride are used as reference points.

Pyroceram, also from Corning, is a glass ceramic composite, and can be used as a radome material [9]. The ceramic crystalline phases are cordierite, magnesium titanate, rutile, and cristobalite, and the remaining 50% is a glass with similar chemistry to the combined crystalline phases. IRBAS is a composite of silicon nitride and barium aluminosilicate from Lockheed Martin (LM in Table 10.3). Brunswick has used aluminum phosphate as the binder in their silicon nitride radome experimentally [10]. Ceralloy make an improved IRBAS (147-31N) and a reaction bonded silicon nitride product (147-01), which can be varied in density to give lower dielectric constants and lighter weight. The properties of Pyroceram, IRBAS, and Ceralloy 147-31N and 147-01, are summarized in Table 10.3.

The dielectric constant for each of those, from room temperature to 1,400°C [11], are shown in Figure 10.5.

The dielectric loss of each over the same range is shown in Figure 10.6. The properties of foam alumina and calcium silicate (wollastonite) are discussed in [12] at 3 GHz and up to 800°C. In addition to adjusting the density of silicon nitride composites, graded composites can be made [13].

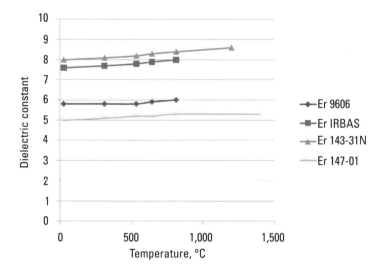

Figure 10.5 Dielectric constant (Y-axis) versus temperature (°C) of high-temperature radome materials.

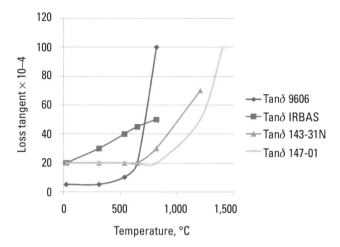

Figure 10.6 Loss tangent (Tanδ × 10^{-4}) versus temperature (X-axis, °C) for high-temperature radome materials.

10.11 Microwave and IR Transparent Radomes

Relatively high–temperature materials are mainly fluoride based. Their microwave and infrared properties are discussed in Chapter 5. For low-temperature applications, high-density (high-polymer chain length) polyethylenes (Chapter 4) have both low dielectric loss and good transparency to infrared at about 10 microns, the equivalent wavelength to human black body radiation temperatures. Polyethylene has narrow infrared absorption bands peaks at 3.43–3.5 microns, 6.7–6.82 microns, and 13.34–13.83 microns, giving it very wide "windows" of transmission. The absorption bands are due to carbon-to-carbon and carbon-to-hydrogen single bond vibration modes. Low microwave loss polyolefin polymers, which have side chains, are cross linked or are co-polymers; all have significantly more modes, as could be expected, and are not suitable for wideband IR transmission. Translucent polymers like PTFE (Teflon) are also not suitable because they contain amorphous and crystalline polymer chains with different refractive indices, causing scattering at optical wavelengths.

10.12   Absorbers for Antennas

Absorbers for general microwave applications are covered in Chapter 3. However, for antennas, foam-based products are used to shape the beam pattern and prevent stray radiation, because of their lighter weight [6–8]. The materials used are foam-based composites of flexible polymers and metal or ceramic polymers. These include natural or nitrile rubbers, polyurethane, silicone, or fluorosilicone polymers, combined with carbonyl iron, iron silicide, ferrite, or carbon absorbing powders dispersed through the polymer. The absorber can be graded for progressive virtually complete absorption, or a mixture of different frequency-selective absorbers can be used. Layers of foam absorber sheets with increasing absorption can be used as an alternative to grading. Many absorbers are used as a Salisbury screen, using a metal backing on the absorber to reflect any radiation back into the absorber. The temperature performance of the absorber is largely dictated by the polymer selected (Chapter 4).

High temperature ceramic foam absorbers can be realized with a silicon carbide absorber in a silicon nitride matrix.

10.13   Phased-Array Antennas

Ferrite-based phase shifter materials from 2 GHz to in excess of 40 GHz are possible using garnet and spinel ferrites (Chapters 1 and 2). Below 2 GHz, weight and size are a major issue for ferrite devices, and although above-resonance devices are possible, much more use is made of switched line phase shifters on dielectric substrates. Microstrip lines with successive phase lengths of 180, 90, 45, and so on can be switched in or out with semiconductor devices. Ceramic tile substrates (Chapters 5 and 6) can be used to reduce size and weight, but contoured laminate can also be used to shape the phased array antenna conformally to the surface of the equipment platform.

References

[1]   Buell, K., D. B. Cruickshank, and K. Sarabandi, "Patch Antenna over the RIS Substrate: A Novel Miniaturized Wideband Planar Antenna Design," *IEEE AP-S*, Vol. 4, 2003, pp. 239–242.

[2] Buell, K., "Development of Engineered Magnetic Materials for Antenna Applications" Ph.D. thesis, University of Michigan, 2005.

[3] Sarantel Ltd., http://www.sarantel.com.

[4] Fiedziuszko, S. J., et al., "Dielectric Materials, Devices and Circuits," *IEEE Trans. MTT*, Vol. 50, No. 3, 2002, pp. 706–720.

[5] Hower, H. T., and S. V. Hoang, "Polyimide Foam-Containing Radomes," U.S. Patent 5662293, 1997.

[6] Emerson and Cuming, http://eccosorb.com.

[7] Laird Technology, http://www.lairdtech.com.

[8] ARC Technology, http://arc-tech.com.

[9] Stookey, S. D., "Method of Making Ceramics and Product Thereof," Corning Inc., U.S. Patent 2920971.

[10] Chase, V. A., and R. L. Copeland, "Fiber Reinforced Ceramics for Electromagnetic Window Applications," *IEEE Trans. Aerospace Supplement*, 1965, pp. 495–500.

[11] Chen, F., Q. Shen, and L. Zhang, "Electromagnetic Optimal Design and Preparation of Broadband Ceramic Radome Material with Graded Porous Structure." *Progress in Electromagnetic Research*, Vol. 105, 2010, pp. 445–461.

[12] Mangels, J., and B. Mikijelj, "Ceramic Radomes for Tactical Missile Systems," Ceradyne Inc. white paper, 2010.

[13] Pentecost, J. L., and P. E. Ritt, "Lightweight Ceramic Materials as High-Frequency Dielectrics," *IRE Trans Component Parts*, 1957, pp. 133–135.

Selected Bibliography

Zurcher, J. F., and F. E. Gardiol, *Broadband Patch Antennas*, Norwood, MA: Artech House, 1995.

11

Tunable Devices

11.1 Introduction

For virtually every microwave resonator or filter application, there is some potential need to tune the device electronically, usually at a distance by appropriate software-controlled commands. The ability to tune is hampered by the need to retain Q and the appropriate impedance, implying a stable but monotonically variable range of dielectric constant or permeability throughout the desired frequency tuning range. Most current solutions use mechanical means of tuning, which tend to maintain Q, but are slow, not easily automated, and can be bulky and expensive.

The methods of tuning, other than those based on semiconducting devices like varactors, can be divided up into magnetic, dielectric, and micro-mechanical. This chapter is largely concerned with magnetic and dielectric tuning.

11.2 Magnetic Tuning

At frequencies in the range 1 to 1,000 MHz, it is possible to magnetically tune the permeability of garnets and spinels with an external DC magnetic field. Bush [1] investigated the change in magnetic spectra of an aluminum-doped YIG with magnetization of about 800 gauss and an undoped nickel spinel. Nickel ferrite does not show the range of permeability required (Figure 11.1 using data from [1]).

The 800-gauss garnet has a desirable range of permeability at field and frequency, although the losses are high below 50 MHz (Figure 11.2). A series of aluminum garnets from 300- to 800-gauss magnetization, measured over 20 to 40 MHz with varying magnetic bias, have lower losses with the lowest magnetization garnets; retaining the range of permeability required a compromise in magnetization (550 gauss) to optimize low loss versus tuning range. This is important because, in the application for an accelerator cavity tuning using a ferrite-loaded coaxial resonator transmission line, the permeability range must be sufficient to tune the cavity over a chosen range of DC field, over the required frequency range [2–4]. Over limited frequencies and DC fields, the Q can exceed 1,000.

11.3 Lumped Element Magnetically Tunable Filters

It has been shown recently [5] that lumped element tunable filters can be magnetically frequency tuned using electrically small toroids integrated into LTCC. This type of toroid can operate up to 700 MHz, but were used as tunable elements in filters from 100 MHz to 200 MHz, with a Q of about 70. Higher-frequency filters using the same

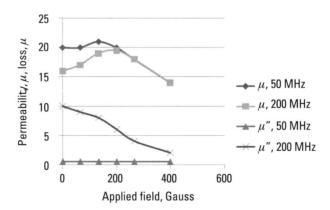

Figure 11.1 Ni spinel permeability (μ) and magnetic loss (μ'') with applied DC field (X-axis) in gauss at 50 and 200 MHz. (*After:* [1].)

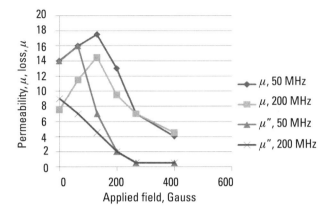

Figure 11.2 The 800-gauss YAlIG permeability (μ) and magnetic loss (μ'') with applied DC field (X-axis) in gauss, at 50 and 200 MHz. (*After:* [1].)

technique should follow, limited only by the minimum size possible that toroids can be made.

11.4 Ferrite Phase Shifters

Tunable ferrite phase shifters are used for beam steering in phased array antennas. Virtually all phase shifters are used below resonance and cover the range of permeability above the low-field loss frequency but below and well away from the low-frequency side of the tail of the ferromagnetic resonance, where the RF permeability is close to unity [6]. Most 360° phase shifters are between 6 and 10 wavelengths long, and therefore need 1 to 1.5 wavelengths of phase shift. Since the propagation constant and therefore the phase is proportional to $Er^{1/2} \times \mu^{1/2}$, then for a fixed Er, μ must be varied magnetically by about 30%, an achievable goal originally derived experimentally by Green and Sandy [7]. They showed that with a suitable choice of material for a given frequency, the magnetic losses were low enough ([8, 9], $\mu'' < 4 \times 10^{-3}$) for most applications, as dielectric losses of most materials are $< 5 \times 10^{-4}$. Ferrite phase shifters from about 2 to more than 40 GHz can be made with insertion losses between 0.5 and 1 dB per 360° of phase shift, depending on the level of fast relaxer doping for power handling. Switching or latching phase shifters using DC current pulses achieve this figure over wide temperature and frequency ranges. Because of hysteresis, resetting pulses are necessary when switching phase settings.

11.5 Magnetically Tunable Microstrip Filters

Magnetically tunable microstrip filters have been fabricated on polycrystalline YIG substrates [9]. The bandwidth measured at −14 dB return loss fell from > 400 MHz at 5.95 GHz to > 200 MHz at 7.37 GHz over a normally biased range of 100 gauss to 1,400 gauss, where losses rose near ferrimagnetic resonance to about 1 dB. The Qs of the two resonators in the filter varied between 1,000 and 400 for one and 200 and 150 over the bias range. Longitudinal bias gave greater tuning range but much faster reducing bandwidth with tuning.

11.5.1 Magnetically Tunable Dielectric Resonator Filters

This approach was extensively investigated by Krupka [10–12]. Using a ferrite disk above, or a ferrite rod through the center of a dielectric TE_{011} resonator (where the H field maximum occurs), the ferrite can be tuned magnetically using the z component of the permeability when an external circumferential (disk) or axial (rod) DC magnetic field is applied. Using the data from [7] and [8], materials of appropriate magnetization were chosen for a number of frequency bands, and a figure of merit (FEM) could be derived experimentally both from the axially magnetized z component of the permeability, and the circumferentially magnetized component. FEM was then $\delta\mu/\mu \times \tan\delta_\mu$, where $\delta\mu$ is the tuning range of the permeability with low magnetic loss, defined as $\tan\delta_\mu$ ($< 10^{-3}$). Some selected data from [10] are included in Table 11.1.

The data appears to show the difficulty in finding low linewidth low magnetization materials (determined by K1/Ms when Ms is low, such that the first-order magnetocrystalline anisotropy tends to be higher) necessary at 2.1 GHz to avoid low field losses. However, at higher frequencies better FEMs are found. Krupka suggests that since the magnetic Q contribution to the overall Q of the resonator/cavity structure is FEM × frequency/2 × maximum frequency shift, this would be of the order of 10,000 for 1% tuning bandwidth for a material with an FEM of 200. Using a cavity 60 mm in diameter and 30

Table 11.1

Figure of Merit for Selected Garnets

Material	$4\pi Ms$ (Gauss)	Frequency (GHz)	μz	Tan$\mu z \times 103$	FEM
YAlIG	550	2.1	0.75	1.09	230
YCaVInIG	1,368	5.2	0.72	1.0	380
YIG	1,751	6.6	0.73	0.9	410
YCaVInIG	1,812	6.6	0.66	0.8	640

Source: [9].

mm height, a high Q, Er 30 (BZT, Chapter 6) TE$_{011}$ dielectric resonator (Chapter 9) of 31-mm diameter and 9.3-mm height and both axially and circumferentially magnetized 550-gauss YAlIG disks, two-pole filters were built at a center frequency of 2.33 GHz with an insertion loss < 0.4 dB, equivalent to a unloaded Q of 8 to 12,000, over a tuning range of 12 MHz. This could be expanded to 28 MHz with an insertion loss of 0.75 dB and corresponding unloaded Q of 3,500. The main drawback was hysteresis.

11.6 Single-Crystal YIG Resonators

Since the discovery of the very low ferrimagnetic line widths of garnet crystals in the 1960s (Chapter 1), single-crystal YIG has been used as a tunable resonator in filters and oscillators. By optical polishing of single-crystal YIG spheres, and suitable crystal alignment, line widths well below 1 Oersted were found at X-band [13]. Because of saturation effects, such spheres increase in line width below about 2.5 GHz [14]. The low spinwave threshold of single-crystal YIG has also been used for limiters [15]. To lower the frequency of operation of single-crystal devices, gallium-doped YIG is used (Chapter 1) in preference to aluminum-doped YIG because of its superior Curie temperature at high doping levels, and narrow line width [16, 17]. Low magnetization garnets like bismuth calcium vanadium garnets (usually referred to as CalVanBIG) also have low single crystal line widths. Lithium spinel single crystals were found to be generally more difficult to make because of lithium volatilization and the need to order crystals at 750°C to obtain linewidths comparable with garnet single crystals.

The range of linear tuning behavior for pure YIG single crystals extends indefinitely above 2.5 GHz because the applied field is simply a function of the gyromagnetic constant (2.8 MHz/Oe), eventually being limited by the available magnetic field in a compact housing to around 20 GHz. The equivalent Qs for line widths in the 0.3 to 0.5 Oersted region is roughly 8,000. Simple loop coupling is used from triplate lines to couple resonators, and filters [18] up to seven sections have been made.

Operation at low temperatures introduces losses from the rapid increase in anisotropy with reducing temperature for both YIG and gallium-doped YIG [16, 17], and the cost of making relatively large pieces of single crystal garnet for structures like low temperature phase shifters. Polycrystalline materials can be made with much lower anisotropy than YIG using indium or zirconium on the octahedral site (Chapter 1) with line widths approaching those of single crystals (Chapter 1, [6]), and [19, 20], which is shown again in Figure 11.3. At

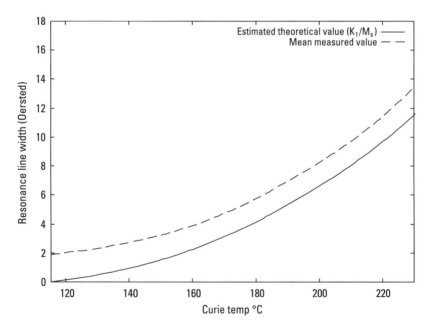

Figure 11.3 Line widths of octahedrally substituted (In, Zr) CaVIGs in the $4\pi Ms$ range 1,000 to 1,600 gauss, versus Curie temperature.

substitutions in the range M = 0.5 to 0.7 in the formula unit in YCaM-VIG, where V is varied to adjust magnetization without affecting Curie temperature, the magnetization can be varied between 1,000 gauss and 1,600 gauss. Ca must be adjusted for the difference between trivalent In and tetravalent Zr, in addition to pentavalent V.

Some indication of the possible line widths from single-crystal materials can be seen in Table 11.2.

Table 11.2 shows that it is possible to achieve single-crystal line-widths of less than 1 Oersted, even for low $4\pi Ms$ materials, implying use down to less than 1 GHz, of which the best is CaVanBIG, because of gradients of the nonmagnetic ion in both Al- and Ga-substituted YIG. This should be compared with results from polycrystalline material shown in Table 11.3. Note that single crystals have to be aligned by referencing crystallographically off their easy axis of magnetization to achieve the minimum line width, whereas polycrystalline materials are essentially isotropic and do not require alignment.

The commercially available materials quoted at their lowest linewidth are usually only supplied specially selected. Values lower than 6 Oe for polycrystalline garnet from commercial companies have been

Table 11.2
Line Width of Single Crystal Garnets and Spinel

Material	Composition	Line Width (Oe)	Frequency GHz	$4\pi Ms$ (Gauss)	Comments
YIG	$Y_3Fe_5O_{12}$	0.3–0.5	9.3	1,780	Flux-grown
YGaIG	$Y_3Fe_{5-x}Ga_xO_{12}$	0.6–1.0	9.3	600	Ga gradients possible (flux)
YAlIG	$Y_3Fe_{5-x}Al_xO_{12}$	<1	X-band		Al in octahedral and tetrahedral sites
CaVanBIG	$Bi_{0.4}Ca_{2.6}$ $V_{1.3}Fe_{3.7}O_{12}$	0.8	1.5	600	Flux
YCaInVIG	$Y_{3-2x}Ca_{2x}$ $In_{0.35}V_xFe_{5-x-0.35}O_{12}$	~1	L- to X-band	0–1,800, x = 1.2-0	In substitution problem
Li spinel	$Li_{0.5}Fe_{2.5}O_4$	~1	X-band	3,500	

Table 11.3
Line Widths of Commercial and Experimental Polycrystalline Garnets

Material	Composition	Line Width (Oe)	Frequency GHz	$4\pi Ms$	Comments
YIG	$Y_3Fe_5O_{12}$	16	9.3	1,780	Fully dense, commercially available
YAlIG	$YFe_{5-x}Al_xO_{12}$	20	9.3	600–1,780, $x = 0.85$ to 0	Fully dense, commercially available
YCaZrVIG	$Y_{3-x-2y}Ca_{ax+2y}$ $Fe_{5-x-y}Zr_xV_yO_{12}$	6 to 10, $x = 0.35$ to 0.4, $y = 0.7$ to 0	9.3	800–1,950	Fully dense, commercially available
YCaInGeIG	$Y_{3-y}Ca_yFe_{5-x-y}$ $In_xGe_yO_{12}$	Low as 1.7, $x = 0.7, y = 0.1$	X-band	465–1,747, $x = 0.4$ to 0.7, $y = 0.1$ to 1.0	Fully dense, Chapter 1, [6]
YCaZrIG	$Y_{3-2x-y}Ca_{2x+y}Zr_y$ $V_xFe_{5-x-y}O_{12}$	Low as 1.8, $y = 0.4, x = 0.4$	X-band	600–1,800, $x = 0.2$ to 0.8 $y = 0.3$ to 0.5	Fully Dense [20]
YCaSnGeVIG	$Y_{3-y-x-2z}Ca_{ay+x+2z}$ $Sn_xGe_yFe_{5-x-y-z}$ V_zO_{12}	Low as 1.9, $y = 0, x = 0.5,$ $z = 0.45$	9.29	0–1,200	Fully dense, [21]

supplied for special applications such as cryogenic (see this Section 11.10) phase shifters [21] and biological sensors. One interesting potential application [22] is the use of submillimeter single-crystal and polycrystalline narrow line width spheres, whose presence can be detected in an RF field remotely with a DC magnet by looking for ferrimagnetic resonance. The local temperature of the single crystal sphere can be determined by the broadening of the anisotropy contribution to the linewidth with decreasing temperature, using the polycrystalline sphere, which does not change significantly, as a reference linewidth peak.

11.7 Epitaxial Thin-Film Magnetically Tuned YIG Devices

Planar devices were found to be possible using gadolinium gallium garnet (GGG) substrates as these are a good lattice constant match for

YIG. When single-crystal YIG was deposited epitaxially, it could then be etched out by photolithography to form disks resonators when biased to ferrimagnetic resonance. The disks were coupled via microstrip or triplate transmission lines. Tunable filters were made initially with bandwidth as low as 14 MHz, which could be magnetically tuned from 2.1 to 2.9 GHz, with an insertion loss as low as 0.5 dB, equivalent to an unloaded Q > 700 [14]. GGG/epitaxial YIG devices were also made for limiters and delay lines [15]. One drawback was paramagnetic losses in GGG at frequencies > 10 GHz [23].

YIG-tuned oscillators and filters using epitaxial garnet film were developed for military and instrumentation applications, mainly by Adam [24]. Bandstop and bandpass YIG/GGG filters laid directly on gallium arsenide have been constructed with high-speed tuning [25].

11.8 Ferroelectric-Tuned Devices

Bulk ferroelectric devices were made using the perovskite (Chapter 6) (Ba,Sr) TiO_3 (usually referred to as BSTO), using the transition temperature range between ferroelectric and dielectric behavior, called the paraelectric region, which can be adjusted using the Ba/Sr ratio. The Curie temperature above which the material is in the desired paraelectric state is defined as that temperature with the highest dielectric constant through the transition region. The dielectric constant falls thereafter with increasing temperature as the paraelectric effect is reduced and the material is essentially a dielectric, with no tunability. For example, at a ratio of 30/70, the dielectric constant of thin films has been reported as 700, with a tanδe of 0.03, and a Curie temperature of $-50°C$ (see Section 11.10). When the Curie temperature reaches room temperature, with a ratio of 55/45, the material is paraelectric and has a dielectric constant of 890 and a tanδe of 0.1 [26]. Other deposition methods give different ranges for the Curie temperature, but most compositions studied lie in the range x = 0.25 to 0.45 in the formula $Ba_x Sr_{1-x} TiO_3$.

Bulk material mixed with MgO were used to make the first experimental devices like phase shifters and tunable filters, using about 60% MgO to reduce the dielectric constant to around 200, and loss tangents into the 0.01 region (Figure 11.4). Bulk $Ba_{0.4} Sr_{0.6} TiO_3$ with

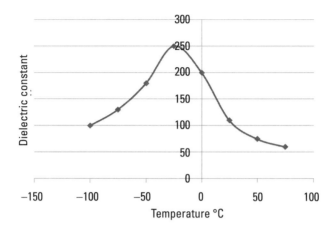

Figure 11.4 Dielectric constant (Y-axis) versus temperature for an idealized BSTO/MgO composition with a Curie temperature of −25°C.

60% MgO has a peak dielectric constant of almost 300 at the Curie temperature, falling to below 100 at about 50°C.

Progressively, however, thin-film techniques such as laser ablation, sputtering, and chemical vapor deposition have replaced bulk methods. BSTO made by these methods can be made very thin, and tuning is measured in volts per micron of film thickness, compatible with voltages used in microwave microstrip active subsystems. BSTO can now be deposited on a wider range of substrates, including high-resistivity semiconductors and alumina, rather than more specialized materials like lanthanum aluminate. Capacitive ranges of more than two have been reported, with reasonable losses.

The main questions surrounding BSTO devices relate to temperature range, an inherent limitation to the technology and similarly, temperature stability. The loss tends to rise with frequency (Figure 11.5). Power and nonlinearity with power parameters are currently being established, because of the power handling, IMD, and harmonic distortion requirements in some systems.

Tunable capacitors using BSTO have become established components, which can be used to make tunable filters and phase shifters. A phase shifter [27] was fabricated from a magnetron sputtered $Ba_{0.65}Sr_{0.35}TiO_3$ film, 0.5 micrometer thick, deposited onto alumina. Using BSTO capacitive tuning, the device was set up to operate over

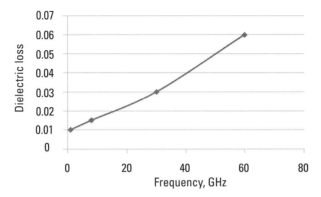

Figure 11.5 Dielectric loss (Y-axis) versus frequency (gigahertz) for a BSTO thin film.

8–10 GHz. The capacitors had a tuning ratio of 2 and loss tangents of 0.03 to 0.05 over the frequency range. Calculated performance suggested a figure of merit of 190° phase shift/dB from the capacitors, but circuit losses dropped this to about 130°/dB. More recently [28], tunable bulk acoustic resonators filters (FBAR) using BSTO on Bragg reflectors have been fabricated to more than 20 GHz. QF products of > 1,000 were obtained. Gevorgian [28] includes a comparison table between different tuning technologies. An abridged version of the data is shown in Table 11.4.

New materials, such as bismuth zinc pyrochlore and silver tantalum niobium oxide are being examined for possible use as paraelectric materials.

11.9 Tunable MEMS Devices

Tunable micron scale devices are possible with micromachined materials (MEMS) using semiconductor device processing technology. Large-scale mechanical devices have low loss and good power linearity because they are insulators or metals with no RF field–dependent magnetic or dielectric properties. When scaled down these properties are largely maintained and these have been used for micron scale mechanical switching. The principle of many of these devices is a cantilevered beam, which moves in response to an electrostatic charge. Although

Table 11.4

Comparison of Tuning Technologies

Technology	Device	Power Consumption	Bias Voltage	Tuning Speed	Q
Semiconductor	GaAs Schottky	< 1 mW	< 5	< 1 nS	200
Semiconductor	Hyper-abrupt varactor	< 1 mW	< 20	< 5 nS	40
Magnetic	YIG tuned	High	Coil current	< 5 mS	>> 3,000
Ferroelectric	Bulk	Negligible	<15K	< 1 mS	> 500
Ferroelectric	Thin film	Negligible	< 30	< 1 nS	> 100
MEMS	Varactor	Negligible	< 50	> 10 mS	< 200
MEMS	Piezo transducer	Negligible	> 100	>100 mS	> 500

Source: [28].

the device can be used as an on-off switch, it is also possible to use it as a continuously variable capacitor in a tunable filter or oscillator [29]. It seems likely these devices will eventually compete with semiconductor-based varactors or switches and with paraelectric devices using BSTO [28], and would combine with SAW and BAW devices on microelectronic scale microwave circuits where high-Q (thousands) filter switching is required. Their main limitation is power handling and possible reliability, given they involve moving parts.

11.10 Low Temperature and Cryogenic Devices

Systems working in space and some devices like parametric amplifiers and high performance filters may need to operate at very low temperatures. This section is concerned with the low temperature properties of materials used in these devices.

11.10.1 Magnetic Materials at Low Temperature

In general it can be said that all magnetic materials increase in magnetization and permeability as the temperature decreases, and that magnetic losses associated with anisotropy and, by extension, fast relaxation processes, also increase. A good example is $Gd_3Fe_5O_{12}$, or GIG. This has a magnetization of > 7,000 gauss near absolute zero [30], as high as any material at room temperature, and more than 10 times its own room temperature value, because of the sublattice compensation. It has a respectable line width between the low temperature side of the compensation point and absolute zero, and has a quite different permeability spectrum at low temperature [31], where it deviates from Snoek's law significantly because of the compensation point. Garnets containing large amounts of magnetic rare earths can be expected to behave in similar ways depending on the sublattice moment and its effect on the other sublattice(s). Most other ferrites have predictable permeability at low temperature, although those containing Co^{+2} vary significantly as the ions positive contribution to anisotropy (Chapter 2) increases with reducing temperature.

Because of the compensation point it is not possible to use Gd-substituted garnets at a low temperature, and even YIG and YAlIG are not used because of the rapidly increasing anisotropy contribution to line width at a low temperature, preference being given to ultra low–line width vanadium garnets doped with indium, tin, or zirconium. Cryogenic devices like phase shifters have been built with these materials with useful results [21].

11.10.2 Dielectrics at Low Temperature

Dielectrics, in general, improve in Q dramatically with reducing temperature, particularly materials like sapphire, where the Q can exceed 10^6, as residual loss mechanisms disappear. Superconducting metal cavities have been used for base station filtering, but in fact comparative performance is possible by simply cooling high-Q materials like BZT and BMT, where Qs of 10^5 are achievable. Two phase materials like $MgTiO_3/CaTiO_3$-based dielectrics improve Q at lower temperatures but because of their nonlinear temperature coefficient, will deviate significantly in dielectric constant and temperature coefficient.

Exceptions to low-temperature high-Q behavior are materials like $SrTiO_3$, which is a good dielectric at room temperature and above, but which becomes ferroelectric at liquid nitrogen temperatures, and, like BSTO, will show ferroelectric then paraelectric behavior between 77K and room temperature.

Rubber- and plastic-based dielectrics will show unusual behavior at a low temperature. PTFE (Teflon) is well known for a phase change near room temperature, with significant dielectric constant and mechanical consequences. Many other polymers undergo second-order phase changes (some are glass transitions) at a low temperature and should be researched before use.

Absorbers also show significant changes with temperature. Magnetic materials depending on low field loss will change because of magnetization and anisotropy increases, which will normally increase absorption if near the low field loss frequency. Water of course becomes ice and in doing so loses its absorption and high dielectric constant. Metal powder and flake-based polymer absorbers will change due to the metal's permeability and conductivity increase with reducing temperature. Those depending on semiconducting mechanisms will lose some absorption if conductivity falls with reduced temperature.

11.10.3 Superconductors at Microwave Frequencies

There are a number of limitations to the use of microwave superconductors. One is the so-called frequency crossover, where, because of surface resistance effects, the effective conductivity reduces as the frequency rises, until it is the same as copper at a crossover frequency. This effect is dependent on the superconductor film quality and is normally well into the millimeter frequency range. The second limitation is the Meissner effect, which limits the RF power handling of the conductor to a critical level. The third is the influence of magnetic fields from switching or phase shifting currents, which will further degrade the superconductor. Nevertheless, a series of cryogenic microstrip ferrite junction (ring resonator) and phase shifters devices were successfully designed, built, and tested at MIT [32].

References

[1] Bush, G. G., "Modification of the Complex Permeability of Garnet and Spinel Ferrites by Application of a Static Magnetic Field," *J. Appl. Phys.*, Vol. 64, No. 10, 1988, pp. 5653–5655.

[2] Smythe, W. R., "Reducing Ferrite Tuner Power Loss by Bias Field Rotation," *IEEE Trans. Nuc. Sci.*, Vol. NS-30, No. 4, 1983, pp. 2173–2175.

[3] Hutcheon, R. M., "A Perpendicular-Biased Ferrite Tuner for the 52 MHz Petra 2 Cavities," *IEEE PAC 1987*, 1987, pp. 1543–1545.

[4] Pivit, E., S. M. Hanna, and J. Keane, "Fast Ferrite Tuner for the BNL Synchrotron Light Source," *IEEE PAC 1991*, 1991, pp. 744–746.

[5] Hopperjans, E. E., and W. J. Chappell, "The Use of High Q Toroid Inductors for LTCC Integrated Tunable VHF Filters," *IEEE IMS 2009*, 2009, p. 905.

[6] Sakiotis, N. G., and D. E. Allen, "Slow-Wave UHF Ferrite Phase Shifters," *J. Appl. Phys.*, Vol. 33, No. 3, 1962, pp. 1265–1266.

[7] Green, J. J., and F. Sandy, "Microwave Characterization of Partially Magnetized Ferrites," *IEEE Trans. MTT*, No. 6, 1974, pp. 641–645.

[8] Green, J. J., et al., "A Catalog of Low Power Loss Parameters and High Power Thresholds for Partially Magnetized Ferrites," *TRECOM-0174-2 Tech. Report*, 1972.

[9] Leon, G., and R. R. Boix, "Tunability and Bandwidth of Microstrip Filters Fabricated on Magnetized Ferrites," *IEEE Microwave Wireless Comp. Letters*, Vol. 14, No. 4, 2004, pp. 171–173.

[10] Krupka, J., A. Abramowicz, and K. Derzakowski, "Magnetically Tunable Filters for Cellular Communications Terminals," *IEEE Trans. MTT*, Vol. 54, No. 6, 2006, pp. 2329–2335.

[11] Krupka, J., "Resonant Modes in Shielded Cylindrical Ferrite and Single Crystal Dielectric Resonator," *IEEE Trans. MTT*, Vol. 37, No. 4, 1989, pp. 691–697.

[12] Krupka, J., A. Abramowicz, and K. Derzakowski., "Magnetically Tunable Dielectric Resonators Operating at Frequencies About 2 GHZ," *J. Appl. Phys., Phys. D*, Vol. 37, 2004, pp. 379–384.

[13] LeCraw, R. C., E. G. Spencer, and G. S. Porter, "Line Width in Yttrium Iron Garnet Single Crystals," *Phys. Rev.*, Vol. 110, 1958, p. 326.

[14] Archer, D. L., W. L. Borigianni, and J. H. Collins, "Magnetically Tunable Microwave Bandstop Filters Using Epitaxial YIG Film Resonators," *J. Appl. Phys.*, Vol. 41, No. 3, 1970, pp. 1359–1360.

[15] Collins, J. H., and G. R. Pulliam, "Microwave Integrated Circuits Utilizing Epitaxial YIG Filters," *Proc. International Conf. on Ferrites (ICF)*, 1970, pp. 487–490.

[16] Spencer, E. G., and R. C. LeCraw, "Line Width Narrowing in Gallium Substituted Yttrium Iron Garnet," *Bull. Amer. Ceram. Soc.*, Vol. 5, No. 2, 1960, p. 57.

[17] Torii, M., et al., "Homogeneity of YGaIG Single Crystal Grown by the FZ Method," *IEEE Trans. Mag.*, Vol.15, No. 6, 1979, pp. 1732–1734.

[18] Carter, P. S., "Magnetically-Tunable Microwave Filters Using Single-Crystal Yttrium-Iron-Garnet Resonators," *MTT IRE Trans.*, Vol. 9, No. 3, May 1961, pp. 252–260.

[19] Machida, Y., et al., "Magnetic Properties and Resonance Linewidths of Zr and Ti-Substituted Ca-V Garnet," *IEEE Trans. Mag.*, Vol. MAG-8, No. 3, 1972, pp. 444–446.

[20] Takamizawa, H., K. I. Yotsuyanagi, and T. Inui, "Polycrystalline Calcium-Vanadium Garnets with Narrow Ferrimagnetic Resonance Linewidth," *IEEE Trans. Mag.*, Vol. MAG-8, No. 3, 1972, pp. 446–449.

[21] Dionne, G. F., et al., "Cryogenic Magnetic Properties of Iron Garnets Diluted with V-Zr, In-V, and In-Al Combinations," *MMM-Intermag Conf.*, BP-10, 2001, p. 60.

[22] Beckman, F. K., et al., "Remote Temperature Sensing in Organic Tissue by Ferrimagnetic Resonance Frequency Measurements," *11th European MW Conf.*, 1981, pp. 433–436.

[23] Adam, J. D., J. H. Collins, and D. B. Cruickshank, "Microwave Losses in GGG," *AIP Conf. Proc.*, Vol. 29, 1976, pp. 643–644.

[24] Adam, J. D., et al., "Ferrite Devices and Materials," *IEEE Trans. MTT*, Vol. 50, No. 3, 2002, pp. 721–737.

[25] Tsai, C. S., et al., "Tunable Wideband Microwave Band-Stop and Band-Pass Filters Using YIG/GGG—GaAs Layer Structures," *IEEE Trans. Mag.*, Vol. 41, No. 10, 2005, pp. 3568–3570.

[26] Malic, B., et al., "Dielectric Properties of (Ba,Sr)TiO3 Thin Films at Microwave Frequencies by Split Post Resonator Technique," *5th Conf. Microwave Mat. Appl.*, abstract O-A03, 2008, p. 72.

[27] Vendik, O. G., "Insertion Loss in Reflection-Type Microwave Phase Shifter Based on Ferroelectric Tunable Capacitor," *IEEE MTT*, Vol. 55, No. 2, 2007, p. 425.

[28] Gevorgian, S., "Ferroelectrics in Agile Microwave Components," *Armenian J. Appl. Phys.*, Vol. 2, No. 1, 2009, pp. 64–70.

[29] Zeigler, V., "RF MEMS Switches Based on a Low-Complexity Technology and Related Aspects of MMIC Integration," *13th GaAs Symposium*, 2005, pp. 289–292.

[30] Pauthenet, R., "Les Proprietes Magnetique des Ferrites d'Yttrium et de Terres Rares de Formule 5Fe2O3.3M2O3," Thesis, University of Grenoble, 1958.

[31] McDuffie, G. E., J. R. Cunningham, and E. E. Anderson, "Initial Permeability Characteristics of Mixed Yttrium Gadolinium Iron Garnets," *J. Appl. Phys.*, Vol. 31 (S), No. 5, 1960, pp. 47–48.

[32] Dionne, G. F., et al., "Superconductivity for Improved Microwave Ferrite Devices," *Lincoln Lab. J.*, Vol. 9 No. 1, 1996, pp. 19–32.

12

Measurement Techniques

12.1 Introduction

Methods of measuring the magnetic or dielectric properties of materials for microwave applications have to meet a number of criteria. They should be relevant to the application, and they should be relatively quick, reproducible, and easy to transfer from the equipment to a database.

In Chapters 8 through 11, we discussed the device properties and how they can be broken down into specific material parameters. In this chapter we will look at the individual parameters and the best way to measure them.

12.2 Dielectric Constant and Loss

Most methods measure both parameters. The main division between methods is the dielectric constant. In general, it is easier to measure low dielectric constants (roughly 3 to 20) and modest Qs or dielectric losses (in the range 0.01 to 0.00001) using high-frequency waveguide cavity perturbation methods, and higher dielectric constants (20 to 100) and Qs (up to 60K) using resonator loaded lower frequency cavities or the resonator coupled to a transmission line.

There are other methods. For example methods using slabs in a waveguide are very accurate but require expensive samples and inconvenient measurement techniques. Similarly, printed resonators on the material as a substrate require complex sample preparation. Test methods for plastic and plastic composite laminates are discussed in Chapter 4 and Section 12.2.5.

A further complication is changes in dielectric properties over the UHF, microwave, and millimetric frequency range. Most loss mechanisms cause the dielectric loss to increase with increasing frequency, resulting in the so-called QF rule for dielectrics. This rule says that the Q × frequency is a constant for a given material. Experience has shown this is only applicable over relatively small frequency ranges for many materials, and should be treated with great caution. However, in general, most materials display relatively flat dielectric properties between the "steps" down in frequency as successive loss mechanisms take effect, with their corresponding absorption peaks (Figure 12.1). In general it is possible to measure well away from those peaks.

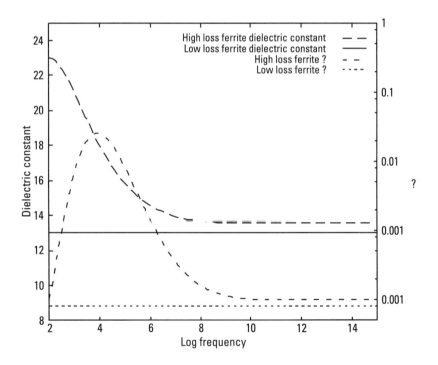

Figure 12.1   Spectrum of low and high loss ferrite dielectric characteristics.

12.2.1 Perturbation Techniques

The most used technique (Figure 12.2) for low dielectric constants (*Er* < 20) is the perturbation method using a small diameter rod in an X-band TE_{103} cavity, [1, 2]. The method depends on the equation

$$Er - 1 = \delta F/2F \times \text{Volume Cavity/Sample Volume} \qquad (12.1)$$

where *Er* is the dielectric constant, *F* is the empty cavity frequency, and δF is the frequency shift produced by the sample.

The change in cavity Q is proportional to the dielectric loss:

$$E'' = \delta(1/Q) \times \text{Volume Cavity/Volume Sample} \times 4 \qquad (12.2)$$

where $\delta(1/Q)$ is the change in cavity Q with and without the sample, and the loss tangent, $\tan\delta e = Er''/Er$.

Since the cavity volume is fixed and the height of the sample and the waveguide narrow wall is the same (1 cm or 0.4"), the equation

Figure 12.2 X-band dielectric measurement.

reduces to being inversely proportional to the sample cross-section area and the frequency shift. This can be measured directly or calculated using the bulk density of the test part and its length if the cross section is not easily determined. For this reason very high–dielectric constants are not measured this way as perturbation theory requires the effective dielectric volume to be small, such that very small diameters are required, with corresponding cross-section measurement errors. The length is normally greater than the narrow wall to allow easy insertion. However, this introduces errors because of leakage or radiation from the cavity and sample. This can be minimized by using a fixed and as short as possible a sample length. The error from the measurement is slight for the dielectric constant but must be carefully adjusted for dielectric loss using variable length samples of a known material.

Almost all observed dielectric losses in ferrite are created by electron hopping when Fe^{+2} ions are surrounded by Fe^{+3} ions. Moderate losses in the range 0.001 to 0.005 do not typically affect the dielectric constant, but higher loss levels tend to produce higher dielectric constants.

A further complication is magnetic loss. In general, a small sample with low magnetic loss can be considered as only entering the E-field of the cavity and therefore behaves only as a dielectric. In practice this is not true of a sample with a practical diameter at X-band (1 mm or 0.04 inch), which has high magnetic loss. For ferrites with $4\pi Ms > 3,000$, $\gamma \times 4\pi Ms$ approaches the measurement frequency of approximately 9 GHz, indicating "low field loss" may be a problem (see Chapter 8). Although this problem can be partly ameliorated by saturating the ferrite with a magnet (without getting close to resonance) a better approach is to use a higher-frequency cavity. Since the highest microwave ferrite has a $4\pi Ms$ of 5,000 gauss, it is convenient to use about 18 GHz and the appropriate waveguide size.

12.2.2 Dielectric Properties Using Dielectric Resonators

The most established method is that based on Courtney [3]. The TE_{011} mode is selected from the mode chart of a shorted dielectric waveguide mode chart. Typically for dielectric constants between 20 and 100 and sample sizes in the range of 12.5-mm (0.5-inch) diameter by 5 mm

(0.2 inch) thick, this will give resonant frequencies in the 3- to 6-GHz range (Figure 12.3). The main issue is accuracy in preparing ground samples, which must be flat to avoid air gaps, and accurately machined to allow precise measurement of dimensions. Courtney also measured the dielectric properties of garnets and obtained very similar results to the ASTM perturbation method. The actual dielectric properties are computed using Hakki and Coleman equations [4] from the measured frequency and Q of the dielectric sample, and its dimensions.

Qs for these types of dielectrics are best measured with appropriate cavity and resonators for the frequency being considered, using the $TE_{01\delta}$ mode and a resonator diameter of about 1/3 of the cavity diameter (Figure 12.4). The cavity contribution to the Q can be calculated from conductivity assumptions. The dielectric loss can then be calculated from 1/Q, the equivalent of an infinitely large cavity. With modern EM simulators it is also possible to derive the dielectric constant from cavity geometries with axial symmetry, although this method is

Figure.12.3 Courtney dielectric resonator test method.

Figure 12.4 $TE_{01\delta}$ test cavity.

not yet universally accepted. Because of potential deviation from the QF rule, the measured frequency should be as close as possible to the anticipated frequency of use.

Humidity is the main variable when Q is measured. The air in an empty cavity will affect the Q when sufficiently humid. In addition, the dielectric test sample can be affected, particularly on its surface, even with a dense, impervious ceramic. Small differences in surface temperature and thermal conductivity between sample and cavity become a problem at high humidity, where condensation will favor a lower temperature, higher conductivity surface. For Qs above about 10K at low microwave frequencies, the humidity is best kept below 40% for reproducible measurements. All high-Q ceramics do degrade under high humidity because of both cavity and dielectric effects, and attention must be paid to these when such conditions are specified in practical applications.

12.2.3 Dielectric Temperature Coefficients

At microwave frequencies, either the dielectric constant coefficient with temperature (Te), or the frequency coefficient with temperature (Tf or τ_f) is used, the latter being more common. Both are expressed in parts per million (ppm) per degree Celsius (°C). They are related by the equation

$$Tf = -\left(1/2\,\text{Te} + \alpha_1\right) \tag{12.3}$$

where α_1 is the coefficient of thermal expansion, also in ppm/°C. The temperature coefficient does not vary significantly over wide frequency ranges but may be nonlinear over temperature, expressed as Tf' in ppm/°C/°C. Both Tf and T'f can be calculated by fitting a second-order polynomial to the frequency change over temperature. For linear materials over temperature, T'f usually needs to be of the order of 0.01 ppm/°C/°C or better. If we assume Tf can be adjusted by changing the composition of the dielectric, which for high-Q materials used in cavities it usually has to be, adjustments may have to be of the order of 0.5 ppm for the most taxing of requirements. The main source of variation in measurement comes from the metal of the measurement cavity. Silver-plated aluminum, brass, copper, and invar have different expansion coefficients, and will give different answers, so reference should be made to the cavity material when comparisons are made. The most common source of error in making the measurement is allowing time at each measurement temperature for the cavity and sample to equilibrate, since these must track with each other. This is a function of the thermal mass of the two and may result in hours rather than minutes for large cavities to equilibrate, unless forced heating or cooling is used.

12.2.4 Low-Frequency Measurements of Dielectric Properties

Useful information about the dielectric properties of materials can be gleaned from low-frequency measurements, using a suitable capacitance bridge or impedance analyzer. A particularly useful approach is to compare the dielectric constant in the range of 100 Hz to 1 MHz (see Figure 12.1). If there is significant dispersion, where the ratio of dielectric constants exceeds 2:1 for 100 Hz:1 MHz, then this is an excellent predictor of dielectric losses at microwave frequencies and can be used to rapidly screen material for potential problems before carrying out more time consuming microwave measurements. All garnet and most spinel ferrites have the same dielectric constant at a low frequency as at microwave frequencies if manufactured correctly, and should not show significant dispersion until approaching far infrared frequencies. Dispersion at such low frequencies is normally an indication of semiconducting phases at grain boundaries in ferrite, and is usually caused by charge imbalance. In CaV garnets, for example, it can indicate an imbalance of +2 to +5 valencies, or of excess Fe^{+2} in all ferrites. For an excellent discussion of dielectric properties of all materials over frequency, see [5].

Measuring very large variations in dielectric constant and loss over wide frequencies is best accomplished with an impedance analyzer and a dielectric test fixture, similar to that used for magnetic behavior (Section 12.3).

12.2.5 Split Resonator Technique

This method [6] has been used extensively for measuring materials in laminate form (Chapter 4), ceramic and ferrite substrates, and thin films coated onto known, characterized substrates. The measurement cavity is cylindrical, and the known, split, high dielectric constant, high-Q resonator used to sandwich the material under test is effectively a TE_{011} composite resonator, which can be analyzed for frequency and Q change with and without the material under test, as shown in Figure 12.5. Coupling is by standard cavity-loop coupling. In the test position, there is no air gap between the resonator halves and the material under test. The dielectric constant and dielectric loss of the material under test can be derived from similar equations to perturbation theory, but with more complex triple integers to derive the volume contribution to cavity frequency and Q.

$$Er = 1 + \delta F / hFK \qquad (12.4)$$

where δF is the frequency difference between the cavity with and without the material under test, h is the thickness of the material under test, and K is an triple integral function of the sample and cavity volume

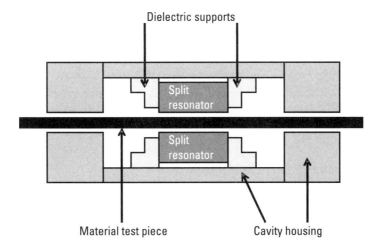

Figure 12.5 Split resonator test fixture (the gaps are normally closed).

and the dielectric constant of the material under test, characteristic of the test setup.

Tanδe is similarly equal to

$$(1/Q \text{ sample} - 1/Q \text{ resonator} - 1/Q \text{ cavity}) \times K' \qquad (12.5)$$

where K' is a corresponding triple integral for the sample and cavity volume and the dielectric loss.

The method is accurate for the low dielectric constant, low loss material if due attention is paid to sample flatness and air gaps in the sandwich.

12.3 Magnetization

Magnetization is defined as the maximum magnetization generated in a material with an external field. In practice, it will depend on the available field, whether the field is DC or AC, and the frequency of the AC field. For that reason the accepted value is normally measured with a DC field, with a term expressing the field dependence at high field values. The so-called saturation magnetization is normally the value seen at fields around 1 tesla, or 10,000 Oersteds of applied field.

For most "soft" ferrites the approach to saturation is essentially flat above a few thousand Oersteds for a spherical specimen and reproducible results are obtained when comparing magnetization with a known traceable standard of the same size and shape, such as nickel measured under the same field conditions. NIST supply standard nickel spheres with known magnetization, temperature, and field characteristics for this purpose.

12.3.1 Vibrating Sample Magnetometer

Although there are many types of magnetometer, we will only consider the vibrating sample magnetometer (VSM) in detail. This approach was first published by Foner [7] and has subsequently been made as a commercial instrument (Figure 12.6). The principle is the AC field generated in a "pick up" coil by the vibrating magnetized specimen in a DC field.

The AC field is a measure of the magnetization in electromagnetic units per gram, or if the volume is known, the saturation magnetization

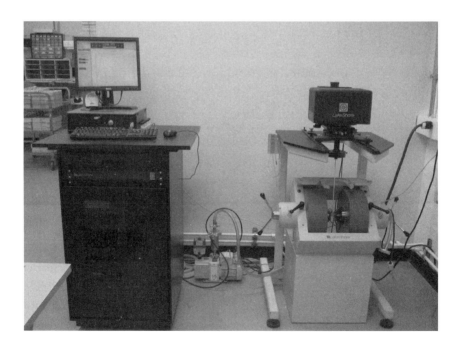

Figure 12.6   Vibrating sample magnetometer.

in gauss. For calibration purposes a nickel sphere of known size is used. The VSM is especially useful for measuring magnetization over temperature because of the ease of heating or cooling the specimen. Hence, the Curie temperature or cryogenic properties of ferrite can readily be determined. A typical magnetization curve is shown for polycrystalline yttrium iron garnet in Figure 12.7, giving the magnetization slope, and by differentiation of the slope, the Curie temperature.

12.3.2 AC Magnetization

Because ferrites show hysteresis, the AC magnetization is measured from the hysteresis loop (Figure 12.8) at the maximum magnetization created by the AC field, known as the induction maximum (Bm) or saturation magnetization (Bs). This is a much lower value than the

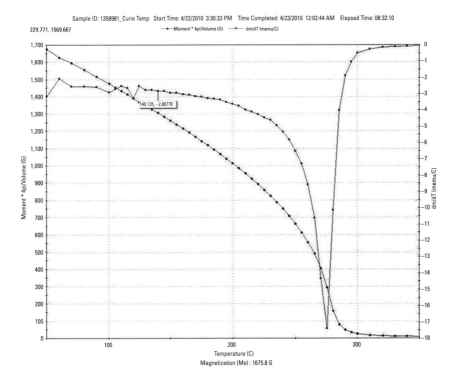

Figure 12.7 Output from VSM showing magnetization versus temperature (the differential of the magnetization versus applied field is used to derive the Curie temperature).

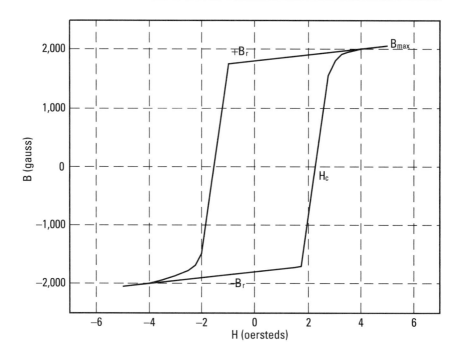

Figure 12.8   BH loop of a typical ferrite.

true saturation magnetization (4πMs) measured by the VSM, as it is measured at typically 10 times the coercive force, in the range of a few Oersteds.

Because of variation in low-field magnetic characteristics, the behavior of Bs and 4πMs with temperature is significantly different, such that low-field latching devices with small latching fields, or below-resonance devices with small DC bias, have different behavior from above-resonance devices. Even in above-resonance devices, the ferrite may not be fully saturated, an important consideration when nonlinear behavior is being addressed.

The hysteresis loop is also used to measure remenance, the residual magnetization when the AC field is removed from a toroidal specimen, and coercive force, the field required to switch the toroid. Measurement conditions are as for Bs. A typical BH measurement system is shown in Figure 12.9.

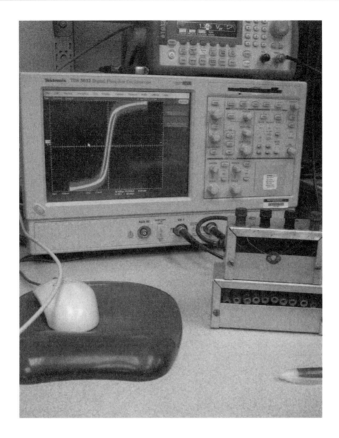

Figure 12.9 BH hysteresis loop measurement.

12.4 Line Width Measurements

Ferrimagnetic line width is the accepted method of measuring magnetic loss, although as discussed in Chapter 1 it does not necessarily indicate off-resonance behavior. It is, however, the basis for the specification of ferrites and a measure of manufacturing reproducibility.

12.4.1 Ferrimagnetic Resonance

Ferrimagnetic resonance is measured by sweeping the DC magnetic field at a fixed frequency, typically with a polished ferrite sphere of about 1 mm (0.04 inch) in diameter. The most common and most reported value for ferrimagnetic resonance is derived from measurements based on an iris-coupled waveguide TE_{106} cavity at X-band and a polished spherical specimen. Other techniques are possible using cross-guide couplers (Figure 12.10) or a crossed microstrip. Essentially the sphere is biased to resonance using an electromagnet, and the DC field half-power points (3 dB) are measured using a Gaussmeter and power meter on each side of resonance. Nuclear magnetic resonance (NMR) is the best gauss meter for this measurement because of its accuracy and internal water (actually a hydrogen proton) reference standard.

The most important aspects of the measurement are, first, sample preparation to give an optical quality spherical surface, using silicon carbide followed by diamond paste processes. Second, control of sample temperature to about 25 +/− 1°C during the measurement is

Figure 12.10   X-band ferrimagnetic resonance line width instrumentation.

necessary as the line width varies considerably with temperature for most ferrites. Third, a single DC field sweep through resonance in one direction avoids hysteresis effects, using suitable software to measure the resonance peak by differentiation of the power against the field curve and computing the 3 dB and if necessary 20-dB points on the resonance curve by nth-order polynomial curve fitting, rather than manual methods that depend on operator setting accuracy, or by assuming Lorenzian behavior, which is not a good approximation for most practical ferrites. Accuracies of about +/− 0.2 Oersted are possible under these circumstances.

The gyromagnetic ratio is the resonance frequency divided by the resonance field and is typically around 2.8 MHz per Oersted. G effective is obtained by dividing the gyromagnetic ratio by 1.4 and is dimensionless.

12.4.2 Spinwave Line Width

This measurement [8] is intended to estimate the power-handling capability of ferrite by measuring the nonlinear point of a spherical sample, usually at X-band. For convenience cavity Q methods are used to make the measurement (Figure 12.11) under reasonable transmitter power. The ferrite under test is biased to saturation below resonance with a DC magnet (omitted in Figure 12.11 for clarity) to create "parallel pumping conditions." Nonlinearity can be measured by direct

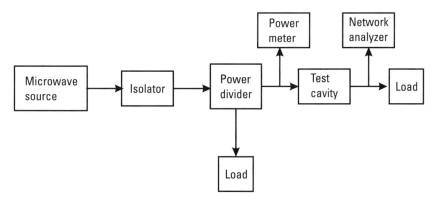

Figure 12.11 Spinwave line width measurement (magnet on test cavity not shown).

observation of cavity CW input and output power levels or more easily by degradation of the leading edge of a pulsed system. Like most nonlinear measurements of this type, accuracy is not good, at best +/− 20%. The method typically can only be used up to about 10-Oersted line width because of transmitter power limitations, and the maximum magnetization of the ferrite is again limited to about 3,000 gauss if measured at X-band because of low field loss. Specimen preparation is not critical but interpretation can be difficult as grain size and porosity can give widely different values for the same composition. The method should always be backed up by physical measurement of grain size and porosity, and indirect measurements such as coercive force, which is more accurate in terms of interpreting grain size effects, particularly of bulk ferrite parts. If fast relaxer dopants are used to adjust spinwave line width, dopant levels can be independently measured by XRF or ICP to quantitatively confirm the line width measurement if doubt exists.

12.4.3 Effective Line Width and Magnetic Losses

As discussed in Chapters 1 and 8, no single measurement has evolved to measure magnetic loss. Both the ferromagnetic and spin wave line width do not accurately predict magnetic losses, particularly when comparing garnets and spinels or ferrites with a different microstructure. It is possible to roughly compare spinels and garnets, which are fully dense, single phase, and with normal grain size (10 to 30 microns) using spinwave line width data in terms of predicting their insertion loss at appropriate frequencies and bias. A further consideration for unmagnetized and magnetized ferrite is the influence of low field loss, which as discussed is both frequency and field dependent. These factors were first considered in ferrite-loaded waveguide [9] and cavity perturbation methods [10] to measure RF permeability and magnetic loss, and these can be used for application-specific comparisons of different ferrite. However, the data should be used with caution for general comparisons, and these measurement methods are not included in detail here.

In reality, a great deal can be inferred about the likely magnetic losses from the absence of fast relaxers and a good microstructure. Below- and above-resonance losses are then likely to be similar in the absence of low field or resonance losses. The real determinant then becomes the resonance line width, which must then be narrow to maximize the potential range of the DC field to give low losses; it is for this reason narrow line width garnets are used wherever possible, especially in above resonance applications.

12.5 Permeability and Magnetic Loss Spectrum

This can be measured over the range 1 MHz to 1.8 GHz with an Agilent HP 4291A/B (Figure 12.12) or AT E4991A Impedance Analyzer, using a toroidal specimen shape in a coaxial transmission line.

The permeability μ is directly proportional to the impedance Z,

$$\mu - 1 = \text{constant} + Z/h \times \log_n (b/c) \qquad (12.6)$$

where b and c are the outer and inner diameter, and h is the height of the toroid. A typical plot is shown in Figure 12.13 for an NiZn ferrite. The plot shows the absorption peaks at 40 MHz and 200 MHz, which can be characteristic of domain wall movement, rotation, or spin resonance. The permeability and magnetic loss (as μ'') can be read off di-

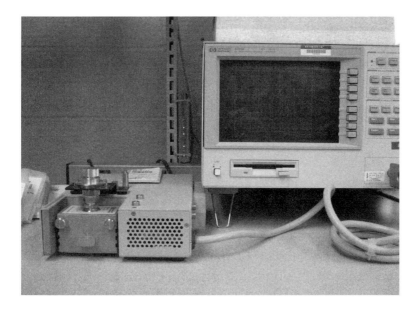

Figure 12.12   Impedance analyzer.

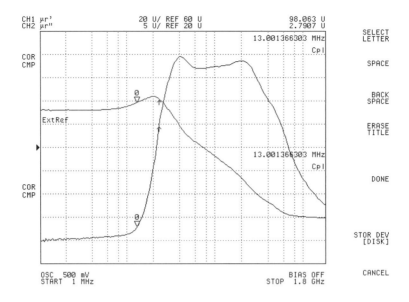

Figure 12.13   Impedance analyzer plot of magnetic spectrum.

rectly at any frequency. It is possible to observe the behavior of ferrites with applied DC fields as the fixture is nonmagnetic.

A similar fixture is used with this type of instrument for dielectric losses, and [11] contains details of both the magnetic and dielectric measurement.

12.6   Intermodulation and Third-Harmonic Distortion Measurement

While this is usually measured on complete devices, there are methods of measuring the contribution from materials. Nishikawa developed a method reported in [12] using a group of three $TE_{01\delta}$ dielectric resonators at different frequencies, in a single line in a cavity, set to give the relationship F3 = 2F2 − F1. By measuring the F1 (about 800 MHz) input power at one of the two inputs, the F2 input power at the other input, and then measuring and comparing the detected power of F1 and F3 in the center of the resonator grouping, the third intermodulation distortion could be measured, provided reflections could be filtered out by F2 and F1 trap filters. The limit of accuracy was reported as −170 dBc. They were able to show substantial variation of dielectric performance in proportion to its dielectric loss.

Miura and Davis [13] used a single ferrite resonator in an enhanced RF field to measure intermodulation distortion when a ferrite is biased by a DC magnetic field. In the below-resonance condition they were able to show the demagnetization shape factor had a significant influence on nonlinear behavior at 2 GHz. Measured values were in the region of −70 dBc, comparable with device measurements, on an 800-gauss aluminum garnet. However, the method has so far not been useful above resonance.

12.7 Density

Perhaps surprisingly critical to predicting electrical and magnetic properties, ceramic densities are measured rigorously in manufacturing. It can be carried out in a number of ways. For dense ferrite or dielectric, the Archimedes water displacement method is used and is accurate for specimens of about 10g and a balance with weighing capability to 1/10th mg. Precautions are typically allowing for normalization of the water temperature to 4°C, and preventing formation of bubbles between the supporting wire and specimen. Detergent can be used to reduce surface tension at interfaces. Accuracies of 0.001 g/cc are possible. Because ferrites' properties are very sensitive to density, this level of accuracy is required. Modern ferrites and dielectrics are fully dense and do not have any open porosity.

Helium or mercury pycnometers can be used for specimens where porosity may exist but are typically only able to measure to accuracies of 0.01 g/cc.

References

[1] ASTM A893-03 , "Standard Test Method for Complex Dielectric Constant of Nonmetallic Magnetic Materials at Microwave Frequencies," 2008.

[2] TransTech, "Test for Dielectric Constant and Loss," Applications note No. 661.

[3]   Courtney, W., "Analysis and Evaluation of a Method of Measuring the Complex Permittivity and Permeability of Microwave Insulators," *IEEE Trans. MTT,* Vol. MTT-18, No. 8, 1970, pp. 476–485.

[4]   Hakki, B., and P. Coleman, "A Dielectric Resonator Method of Measuring Inductive Capacities in the Millimetric Range," *IRE Trans. MTT-8,* July 1960, pp. 402–410.

[5]   Von Hippel, A., *Dielectric Materials and Applications,* Cambridge, MA: MIT Press, pp. 47–139.

[6]   Krupka, J., "Complex Permittivity Measurements with Split Post Resonator," *IEEE IMS*, WS07, 2004.

[7]   Foner, S., "Vibrating Sample Magnetometer," *Rev. Sci. Instr.*, Vol. 30, No. 7, 1959, pp. 548–556.

[8]   Trans-Tech, "Test for Spin Wave Line Width," Applications note No 665.

[9]   Green, J. J., "A Catalog of Low Power Loss Parameters and High Power Thresholds for Partially Magnetized Ferrites," *IEEE Trans. MTT,* Vol. MTT-22, No. 5 1974, pp. 641–645.

[10] Rado, G. T., "Magnetic Spectra of Ferrites," *Rev. Mod. Phys.,* Vol. 25, 1953, p. 81.

[11] Agilent Technologies E4991A Impedance Analyzer, Material Measurement A-002, 2009, p. 17.

[12] Tamura, H., et al., "Third Harmonic Distortion of Dielectric Resonator Materials," *Jap. J. Appl. Phys.,* Vol. 28, No. 12, 1989, pp. 2528–2531.

[13] Miura, T., and L. E. Davis, "Evaluation of Intermodulation Distortion of a Ferrite Element by the Two-Tone Method," *IEEE Trans. MTT,* Vol. 57, No. 6, 2009, pp. 1500–1507.

About the Author

David B. Cruickshank is currently the director of engineering emeritus at Trans-Tech Inc., a subsidiary of Skyworks Solutions Inc., where he is now approaching retirement. He graduated with a degree in physical chemistry from the University of Edinburgh in Scotland. Mr. Cruickshank joined Ferranti Radar in Edinburgh as a staff engineer working on microwave materials, where he was introduced to microwave engineering through his training there and at Napier University. He then joined Microwave and Electronic Systems Ltd (MESL) as a materials laboratory manager. When the company was acquired by RACAL, as a member of the board of Racal MESL, he served as the engineering director; his engineering team won the Queen's Award for Technical Innovation. Mr. Cruickshank later served as the general manager of the Southern Division of the company in London, England. He graduated with a degree in management studies from West Lothian College, with a distinction in business systems, during that time.

Mr. Cruickshank joined Trans-Tech in Maryland as the director of new products and later became the engineering director, working on materials and components using magnetic and dielectric materials. He also served as the first chairman of the Technical Steering Committee at Skyworks. He has authored numerous invited papers and has been awarded several patents in the field of microwave materials during his time in the United States.

Index

Recent Titles in the Artech House Microwave Library

RF Design Guide: Systems, Circuits, and Equations, Peter Vizmuller

RF Measurements of Die and Packages, Scott A. Wartenberg

The RF and Microwave Circuit Design Handbook, Stephen A. Maas

RF and Microwave Coupled-Line Circuits, Rajesh Mongia, Inder Bahl, and Prakash Bhartia

RF and Microwave Oscillator Design, Michal Odyniec, editor

RF Power Amplifiers for Wireless Communications, Second Edition, Steve C. Cripps

RF Systems, Components, and Circuits Handbook, Ferril A. Losee

The Six-Port Technique with Microwave and Wireless Applications, Fadhel M. Ghannouchi and Abbas Mohammadi

Solid-State Microwave High-Power Amplifiers, Franco Sechi and Marina Bujatti

Stability Analysis of Nonlinear Microwave Circuits, Almudena Suárez and Raymond Quéré

Substrate Noise Coupling in Analog/RF Circuits, Stephane Bronckers, Geert Van der Plas, Gerd Vandersteen, and Yves Rolain

System-in-Package RF Design and Applications, Michael P. Gaynor

TRAVIS 2.0: Transmission Line Visualization Software and User's Guide, Version 2.0, Robert G. Kaires and Barton T. Hickman

Understanding Microwave Heating Cavities, Tse V. Chow Ting Chan and Howard C. Reader

For further information on these and other Artech House titles, including previously considered out-of-print books now available through our In-Print- Forever® (IPF®) program, contact:

Artech House Publishers	Artech House Books
685 Canton Street	16 Sussex Street
Norwood, MA 02062	London SW1V 4RW UK
Phone: 781-769-9750	Phone: +44 (0)20 7596 8750
Fax: 781-769-6334	Fax: +44 (0)20 7630 0166
e-mail: artech@artechhouse.com	e-mail: artech-uk@artechhouse.com

Find us on the World Wide Web at: www.artechhouse.com